EXPANDING HERMENEUTICS

Northwestern University
Studies in Phenomenology
and
Existential Philosophy

EXPANDING HERMENEUTICS

Visualism in Science

Don Ihde

Northwestern University Press
Evanston, Illinois

Northwestern University Press
625 Colfax Street
Evanston, Illinois 60208-4210

Printed in the United States of America

ISBN 0–8101-1605-7 (cloth)
ISBN 0–8101-1606-5 (paper)

Library of Congress Cataloging-in-Publication Data

Ihde, Don, 1934–
Expanding hermeneutics : visualism in science / Don Ihde.
p. cm. — (Northwestern University studies in phenomenology and existential philosophy)
Includes index.
ISBN 0-8101-1605-7 (cloth : alk. paper).—ISBN 0-8101-1606-5 (pbk. : alk. paper)
1. Science—Methodology. 2. Science—Philosophy. 3. Hermeneutics. I. Title. II. Series: Northwestern University studies in phenomenology & existential philosophy.
Q175.I365 1999
502.8—dc21 98-47390
CIP

The paper used in this publication meets the minimum requirements of the American National Standard for Information Sciences—Permanence of Paper for Printed Library Materials, ANSI Z39.48–1984.

This one is for my science colleagues at Stony Brook—
thirty years of conversations, conflicts, challenges,
and even entertainment.

Contents

Acknowledgments

The following chapters have been previously published and are reprinted here with the permissions of the publishers noted:

Chapter 1, "Interpreting Hermeneutics," originally appeared in *Man and World* 13 (1980): 325–43.

Chapter 2, "Language and Two Phenomenologies," appeared in the *Southern Journal of Philosophy* 8, no. 4 (1970): 399–408.

Chapter 4, "Whole Earth Measurements," appeared in *Philosophy Today* 41, no. 1 (spring 1997): 128–34.

Chapter 5, "Singing the World," appeared in *Horizons of the Flesh*, ed. Garth Gillan (Carbondale, Ill., 1973).

Chapter 6, "Paul Ricoeur's Place in the Hermeneutic Tradition," appeared in *The Philosophy of Paul Ricoeur, The Library of Living Philosophers*, vol. 22, ed. L. Hahn (Chicago, 1995).

Chapter 7, "This Is Not a 'Text,' or, Do We 'Read' Images?" appeared in *Philosophy Today* 40, no. 1 (spring 1996): 125–31.

Chapter 9, "Response to Rorty," appeared in my *Consequences of Phenomenology* (Albany, 1986).

Chapter 10, "Why Not Science Critics?" appeared in *International Studies in Philosophy* 29, no. 1 (1996): 45–54.

The remaining chapters, although sometimes presented as papers, have not been previously published. Part 4, which I call a minimonograph, is entirely new.

I also wish to acknowledge the help and constructive criticism of Robert Crease and Marshall Spector, colleagues at Stony Brook, and the conversations and enlightenment provided by so many of my science colleagues, including John Marburger in physics, Fred Walters and Deane Petersen in astronomy, Donald Harrington of the medical school, Arie Kaufman of computer science, and so many others. Some of the chapters were read and discussed by the participants of my postdoctoral seminar: Sung Dong Kim, Monique Riphagen, and Ken Yip. Indirectly, the colleagues of the ISHS should be acknowledged for the inspiration of this project. Similarly, the philosophers of science in Scandinavia, Holland, and Germany who have followed the program in instrumental realism contributed to this background. Evan Selinger also helped by doing the index and detecting last-minute errors. And, because the manuscript was prepared during both a change-over of computers and software and several moves of offices, special thanks go to Linda Einhorn-Ihde at home and Letitia Dunn and Virginia Massaro at the office. My son, Mark, always provides inspiration with his curiosity and often sage insights, and my wife, Linda, provides stimulation when I sometimes get frustrated at the nitty-gritty details of writing, and she persuades me to update my often sedimented skills.

Introduction: Expanding Hermeneutics

Bruno Latour begins his *We Have Never Been Modern* with "The Proliferation of Hybrids." Donna Haraway proclaims this to be the time of "Cyborgs" in her *Simians, Cyborgs and Women.* In Latour's case, hybrids proliferate because modernity, by making a strong distinction between things of nature and things of culture, ends up with hybrids which are things between or both of nature and culture. In Haraway's case cyborgs are similar to hybrids, as mixtures of humans and technologies or even other species. All of this ambiguity, this confusion, this bricolage occurs at the juncture where we all stand: the juncture between modernity and what follows, most often now called postmodernity.

What makes for the ambiguity may well be that our framework, forged by modernity, which has served long with its nature-culture distinction, is breaking down, or at least has become shaky. It still frames much of current debate: the various forms of nature-nurture arguments about intelligence, criminal behavior, aggressiveness, and the like are nature-culture framed. But the borders are not clear and may not be borders at all.

Variations of these same modern distinctions have colored our discussions of technologies as well. Here the nature-culture distinction is echoed in the presumed difference between "natural" objects and "artificial" ones, the latter presumably being "technological." From the very beginnings of my musings on technologies I was either skeptical or negative toward this perspective on the range of our material "culture." At first, it was with respect to instruments—since extended to more types of technologies—that I disagreed with both the utopian and dystopian thinkers who both tend to reify technology. Instead, technologies, particularly instruments, I have argued, should be regarded as means by which our perceptions and our wider experience are modified and transformed. In effect, this blurs or rejects any hard nature-culture framework.

Expanding Hermeneutics is, with respect to this juncture of modernity and its next transformation, a sequel to my earlier *Postphenomenology*, which appeared in this series. Here, too, I address the juncture of modernity and postmodernity, but with a focus upon hermeneutics, and particularly hermeneutics and science. This may seem to some to be an odd pairing: what has hermeneutics to do with science? If the image of the hermeneut is that of a literary critic or analyst reading, criticizing, deconstructing a "text," and of the scientist imaged as the investigator—or, as the collaborator with late capitalistic manipulator—of nature, then the very pairing might seem suspect. But, if hermeneutics most broadly construed is *interpretive activity*, then this activity in some sense must span both disciplinary matrices.

As a book, this is a "hybrid" in that it combines a group of related essays on hermeneutics and hermeneutic issues (parts 1 through 3) with a kind of *minimonograph* (part 4) on a program to reframe our understanding of much science praxis in hermeneutic terms.

I begin, in part 1, with a broad view of the history and development of hermeneutics, drawing from some now fairly old essays on hermeneutics as they relate to both linguistic (and textual) forms of interpretation and perceptual activity as a kind of bodily interpretive activity. But then I turn to some very recent work in which I begin to relate hermeneutics to both the philosophy of technology (chapter 3) and developments in science (chapter 4). Hermeneutics in this context may seem less familiar.

To the conversations in contemporary philosophy, which include issues of hermeneutics with respect to language and perception, I now add *images* as a particularly interesting area of concern for this approach, which takes up both "continentals," as Euro-American philosophers are popularly called, and "analytics," as Anglo-American philosophers are sometimes called. Merleau-Ponty and Paul Ricoeur play prominent roles here among "continentals," along with Derek Parfit and Richard Rorty for the "analytics." The issues which arise in the debates and dialogues between these two styles of philosophizing may be seen to revolve around what is placed under constraints (styles of fictions, degrees of relativity, and languages) and what is left unrestrained. In effect, I argue that contrary to the myth that analytic philosophies are "rigorous," I find that the lack of existential constraint allows all sorts of odd and unjustified phenomena to slip into these philosophies. And, again, I conclude each part with a look at the contemporary role of philosophy vis-à-vis science (chapters 7 and 10).

Part 4, "Expanding Hermeneutics," is the minimonograph. It is a much more programmatic and even partially systematic attempt to identify the convergence of hermeneutics with science, to identify the hermeneutic dimensions of science praxis already extant, and to hint at a reframing of how we should understand science in operation, with suggestions toward an even richer future of inquiry. This minimonograph is a more concentrated effort to synthesize elements of recent separate essays which arose from the founding of the International Society for Hermeneutics and Science in 1993 and which are now published in its proceedings.[1]

That first meeting in Hungary, referred to in chapter 3, was a culture shock which sent me scurrying to rethink and reformulate what I had been taking for granted. Something like this happened much earlier with respect to the philosophy of technology, in 1982. Giving a faculty development seminar to a consortium of Colombian universities in Bogota, I had come prepared with presentations on the interrelations between science and technology—virtually a required issue in the North American philosophical context—only to discover that my audience held a radically different perspective upon "technoscience." On the spot, I had to reconceive my approach, and later I continued to learn from this shock. I mention this earlier culture shock because it stimulated a long line of thinking about science and technology along intercultural and political lines, which helped to enrich the notion of contemporary science as *technoscience.*

There, however, I was the "conservative" who came with a kind of ignorance concerning the cultural impact (including the political) which was better discerned by the Colombians. The meeting in Hungary was just the reverse. In this case I quickly sensed that I, and a small group of American and Eastern European colleagues, was the "radical." We were clearly post-Diltheyan in contrast to the strong and outspoken positions taken by the others, who maintained that hermeneutics and science might meet but would remain separate (along Diltheyan lines). Not believing that, I was again forced to rethink and reformulate positions in a stronger way than before. *Expanding Hermeneutics* is primarily the result of that philosophical culture shock.

The overarching aim here is to argue that we have often misconstrued what science is and how it operates because, in part, we have for so long ceded the interpretation of science to forms of positivism. In what I call the "H-P Binary"—the contestation between hermeneutics and positivism—hermeneutics first finds itself divorced from the sciences, and then by its own historical proponents made semiautonomous with respect to its interpretive activities in such a way that positivism simply became the standard for framing the understanding of the sciences. What I call the "P-H tradition"—the phenomenological version of hermeneutics—often itself simply accepted this binary, and until recently tended to ignore attempts to enter the domains of science praxis and the understanding of same.

Science, as an institution and a social entity, is, of course, a creation of modernity. It retains much of modernity within its self-understanding, its rhetoric, and its politics. Many scientists today have even entered a contest with "deconstructionists and postmodernists" in a polemic fashion. Several conferences on these themes have been held, with subsequent anthologies published, such as *The Flight from Science and Reason* (1997). (A review of this book in *Science* argued that many of the scientists writing actually showed their own versions of postmodern trends in the process. Some of the scientists so characterized later responded with outrage to the reviewer, Paul Forman, in the "letters.")[2] This same contest is visible in the Alan Sokal affair involving the journal *Social Text*, in which Sokal, a physicist, spoofed postmodernism—which he also lumps with hermeneutics—by "falsely," as he claims, showing science to be culturally relative.

As amusing as these culture wars may be, I am not directly entering these as such. While I do hold that some science praxis is "postmodern" in the sense that it adapts a kind of both relativity and multiperspectivalism in its operations, most scientists' self-understanding remains within the limits of modernism. But that science is culturally embedded and that science as institution functions according to more or less well-understood social structuring would be denied only by an Enlightenment "reactionary" today.

The focus here, however, is upon hermeneutics. I argue that much of science praxis *is* functionally hermeneutic, but only understandably so if certain modifications are made both to the P-H tradition and to the understanding of science as an embodied *technoscience*, the instrument-embodied science of the contemporary world. I try to isolate the features of a *visual hermeneutics* which operates as the "proof" and the demonstration which supports the production and construction of scientific knowledge.

These are themes which many will recognize from previous work. Yet the twist here is different. It is only recently that I began to understand how unified and cross-disciplinary the hermeneutic dimension of science is—as a visual praxis. What has emerged, largely since midcentury, is a set of instrumental practices which converge upon similar approaches in many, not just a few leading, sciences. From physical anthropology, to astronomy, to the medical sciences (particularly what are now even being called the "imaging sciences"), to earth science and oceanography, the same patterns of multiple instrumentation, multivariant instrumental perspectives, and the like have emerged as the standards of science's depiction of the natural world.

Expanding Hermeneutics leaves much unsaid and undone. I do not enter here the longer issues of changes in epistemology which must accompany and result from these new practices or from this first attempt to reframe our understanding of science. Nor do I take up the hard task of outlining the *critical* dimensions which also necessarily surround all interpretive activity. I also withhold discussion of computer modeling—although I hold that computers are another form of hermeneutic device—which plays equally and increasingly important roles in science's visualism.

As I reread the text for this book, I became quite aware that this first foray carries a tone which could be taken to be too positive, too uncritical, too admiring of the ways in which contemporary science has filled our world with its plethora of objects of which we were previously unaware. Yet the opposite direction bothers me even more: it is my firm belief that were science to eliminate its instrumentation (its technologies which embody its inquiries), to return to the "purer" speculations of earlier and more ancient thought, we would soon be reduced to a Democritean state in which "in principle" we could not perceive the objects of science. We would be reduced to speculations about the shapes and interconnections of the atoms of Democritus but know that we could never get "to the things themselves." Of course, there are always holdouts, and these usually are found among physicists. Today those who want to hold to imperceptibility belong to the quantum mechanicians who often claim that the spooky parts of quantum phenomena cannot be visualized, but are understood only through mathematics—echoing Galilean metaphysics, not Galilean practice. This is not something new: to the contrary, the trajectory toward more "textlike" hermeneutics remains within science itself. Some scientists do not like "pictures" and prefer formulas. Others recogize the value of the "aha" quality of getting a depiction. Here is a precise counterpart to the tension between the "textualists" among postmodern critical theory and phenomenological perceptualist hermeneuts as found in the humanities.

While I believe this, I also realized that by using very recent sources (mostly since 1995), from both books and science magazines (and, although not cited, from my discussions with science colleagues), I had inadvertently reflected and carried over precisely the rhetorical tone of those sources. One can detect this positivity in the quotations themselves. Far from being either philosophically skeptical or humanistically dystopian about new developments, science sources tend to be more than abstractly anonymous—they are positive. I have occasionally tried to rebalance this tendency by citing negative or "revenge" effects.

What I am claiming, in this venture, is that the visual hermeneutics of contemporary science provides a quite different scenario concerning science—and it is that scenario which I have teased out in this approach. Science has a way of seeing which I claim is a visual hermeneutic, and it is that dimension of its activity which is forefronted here.

This hermeneutic, moreover, functions within the history of science not unlike the earlier process that occurred in literate cultures, which much more gradually embedded the "technologies" of writing which, in turn, transformed human languages and the perceptions of the lifeworlds of those who lived within literate worlds. Nor do I think we have reached the end of this transformation.

PART 1

INTERPRETING HERMENEUTICS

Hermeneutics has a history as long as philosophy itself. Aristotle's *Peri hermeneias*, or "concerning interpretation," makes it part of the philosophical canon. This is not to say that it has always been so remembered by philosophers. In the early seventies, I was on the Eastern Division Program Planning Committee, and I suggested that an invited symposium on hermeneutics be offered with Paul Ricoeur as lead symposiast. One of my colleagues, antagonistically, asked, "What's hermeneutics? And why is it interesting?" The symposium did occur with a very large audience. But the resistance continued, if now in a form which must recognize it by name, when years later Adolph Grunbaum lent his presidential address to attacking "the hermeneuts."

In this first part, I begin with a brief history of hermeneutics within European philosophy, at first taking it to its twentieth-century bond to phenomenology. *Interpretive activity*, as I prefer to translate *hermeneuein*, will in this context necessarily have to deal with the language phenomena (speech through writing and reading), but also with perceptual phenomena (sensory interpretive activity).

What I take to be the "expansion" of hermeneutics in this book, however, has to do with taking interpretive activity into "*technoscience*," a term for technology and science coined in the sixties by Gaston Bachelard. Within this context, the development of philosophies of technology, which precisely because they have not been subsumed under the myths of value neutrality and ideality which applied for so long in the philosophies of science, could be seen in more obvious social-cultural and thus hermeneutic terms. Yet, when the turn is made to late twentieth-century science, something begins to gel which calls for a new hermeneutics—the emergence of a unified and specialized mode of *thing interpretation* through imaging instruments. Hermeneutics, I contend, must now include this dimension of interpretive activity.

1

Interpreting Hermeneutics: Origins, Developments, and Prospects

1. The Origins of Hermeneutics

The root word for hermeneutics is the Greek verb *hermeneuein,* which means simply in its most general meaning *to interpret.* But definitions are abstract, and to take note of the more radical origins calls for locating the concrete figures and their role for the understanding of sources. A *source,* as the French retains the sense, is the springhead, the origin from the earth itself of clear and clean water. So with *hermeneuein.* Its concrete references, as Heidegger pointed out, ultimately point to *Hermes,* the winged messenger of the gods, and, in earthly imitation, to *hermios,* who is the priest who interprets the sayings of the Oracle of Delphi, the oracle from whom Socrates claimed authority for his mission. Thus like almost every persistent and important philosophical problem in the West, hermeneutics can be traced back to the Greeks and in particular to the rise of Greek philosophy.

Hermes is the messenger of the gods, he who brings a *word* from the realm of the wordless; *hermeios* brings the word from the Oracle—*hermeneuein* is primordial interpretation, the bringing into word of what was previously not yet word. Hermeneutics is the most primitive sense of "to say." And from this coming to birth of word, of language, its derived meanings of explaining as in bringing to understand, and translating, as in making a foreign tongue or meaning familiar in one's own tongue, arise. By the time these root sources become philosophical in the self-conscious sense of philosophical theorizing, they mean the science or art of interpretation as per Aristotle's *Peri hermeneias,* concerning interpretation, which appears in the *Organon* along with logic, rhetoric, and the analysis of all possible types of human significant utterance.

But just as most of our persistent philosophical problems have one source in Greek culture and the rise of philosophy, there is also another classical and ancient root, the Hebraic or biblical tradition. Once Greek culture became permeated, in the postclassical era, with the influx of both the Jewish and Christian traditions, hermeneutics began to take on a different significance. The biblical culture is a culture of *word,* of *Word made flesh* as the Incarnation was thought of, but more specifically of word as Word of God as expressed in a *text.* The record of the primordial word was the Bible. But the Bible was never self-evident, nor did it exist without a supporting historical context—it called for *interpretation.* To interpret, now the interpretation of a text, implies hermeneutics. And just as with the Greeks the concrete origins in the imagery of messages from the gods became the science of interpretation, so in the post-Hellenic era hermeneutics became the science of the method of interpreting Scripture. (Exegesis was the act of interpretation in the common sense, now, while hermeneutics becomes the narrower and more theoretical "philosophy" of textual interpretation.) This meaning and the distinction which bears upon hermeneutics was amplified again at the time of the Reformation and following. No longer accepting long implicit traditions of taken-for-granted interpretation, Protestantism became self-conscious about interpretation as presumed "original" meanings were sought. Hermeneutics became the recovery of lost origins, a descent through layers of interpretation toward a pristine interpretation, the sense of "original" or primitive Christianity. Thus from the sixteenth century onward, hermeneutics becomes an essential theological enterprise.

At this point it is important to note that both a significant difference and a significant similarity between the meanings of hermeneutics as they occur in the two root sources of our civilization exist. The difference which Christian hermeneutics introduced was that hermeneutics was conceived of *narrowly,* as the theory of interpretation of texts and in particular sacred texts, and with respect to Word which had already become word. Although the word of the Bible *pointed to* the primal event, Word made flesh, it itself was already word, and thus interpretation was a process *in media res,* in the midst of word.

Nevertheless, in this narrower meaning of hermeneutics, a structural similarity with its Greek sense remained. Hermeneutics had as its task the interpretation of that which was primordial, the coming into being of Word, the *event* of meaning. Thus biblical interpretation was "archeological" in that interpretation digs back through layers of significance, ever seeking some shining revelation of meaning—a word from the gods. Hermeneutics remained under the aura of Hermes, a messenger of the sacred.[1]

But the origins of hermeneutics do not cease with the melding of the two classical sources, Greek and biblical. In modern times there has been a growing internal modification within Western culture which is all too familiar to us: the rise of the *scientific era*. We are all vaguely, but strongly, aware that what counts today as truth is affected in some way or other by scientific modes of thinking. That is the case whether we are naively positivist and believe truth to belong solely to logical and empirical realms, or whether we are the most strongly romantic and believe that truth is precisely that which lies entirely outside methodological verification.

This internal modification within Western culture is, in part, an amplification of tendencies which themselves also go back to the rise of Greek philosophy. The development of the "distance" of the theoretical attitude, the valuing as true only that which is universal, lawful, and orderly, and the systematic search for an ultimate coherence—all were part of the original philosophical impetus. Later linked to the materialism of Democritus, Epicurus, and Lucretius, this march of scientific thought branched outward from the Renaissance through the Enlightenment onward to extension from one domain to another until it reached the realm of hermeneutics as well.

By the nineteenth century, scientific attitudes and methods had become partially victorious even with respect to the Bible. Although the Bible might be sacred, it must be interpreted in terms of a theory which would apply to not just sacred texts, but *any* texts. So-called "higher criticism" called for the total literary, historical, and humanistic examination of all ancient texts. At the same time, the rise of the scientific era was also accompanied by another tendency, the tendency to expand the notion of a text. Earlier the Renaissance and early Modern era saw much in the metaphor of a text in the concept of the "Book of Nature" which scientific investigation must learn to "read." Bacon's early endeavors were thus something of an early general hermeneutics of nature. By the nineteenth century the rationalist and scientific revolution was all but final in its victory. The narrower sense of hermeneutics as a specifically theological task had to come to grips with the wider universal claims of scientific criticism and interpretation. It is at this junction that the contemporary sense of hermeneutics begins to take its shape. The two figures probably most responsible for this reshaping were Friedrich Schleiermacher and Wilhelm Dilthey. Without going into detail about what they theorized, it is probably sufficient to note that both returned the sense of hermeneutics to its more general ancient philosophical sense, "to interpret," while also giving hermeneutics something of a new specific shape.

Schleiermacher's answer to the challenge of scientific criticism and thinking was basically to *accommodate* it into the theological task itself: biblical hermeneutics must be transformed into a *general* hermeneutics, universal in scope and valid for the wider problems of interpretation. With Schleiermacher, hermeneutics became the *philosophical* science of understanding itself. The principles of hermeneutics were to be understood as basic to any kind of textual and historical interpretation whatsoever, and biblical interpretation became but one instance of an area of interpretation. Hermeneutics, with Schleiermacher, became humanistic, but it also became broad in its ancient sense as the art or science of interpretation as such. However, in the humanistic direction, hermeneutics was now linked to the problem of human and historical understanding.

Dilthey followed Schleiermacher in a somewhat more specific and less theological direction. He saw hermeneutics as the possible foundation for a science of human and historical dimensions, what we would today call the behavioral and social sciences or human sciences. As a science of historical human understanding, hermeneutics was to enunciate the principles which would differentiate the human or social sciences from the natural sciences. And thus the debate between "understanding" and "explanation" was entered.

The emerging contemporary sense of hermeneutics, then, is one which refers particularly to the interpretation of the human, and one might say that by the end of the nineteenth century hermeneutics was already the latent *existential* science.

Were one to follow linearly the progression of the history of the use and transformation of hermeneutics, the next logical step would be to turn directly to contemporary hermeneutic philosophers, but to do so would sidestep the single largest *indirect* development underlying contemporary hermeneutics: the development of *phenomenology*, and in particular the phenomenology of Edmund Husserl. A number of perennial philosophical issues arise at this juncture.

2. Phenomenology and the Transformation of Hermeneutics

Husserl's problems and their sources, at first, seem rather far removed from those of either classical or early contemporary hermeneutics. Rather, Husserl's concerns for a radical reformulation of the sciences arose within the traditions of Modern rationalism and empiricism. The ideal—the dream—of Modern philosophy had been for a truly radical beginning, a search for some absolute grounding from which to build, step by step, a certain and universal science. And whether this was the innate, clear, and distinct ideas of the cogito of Descartes or the simple ideas of Locke at the origins of empiricism, this search was also Husserl's. Husserl's phenomenology was the search for yet another absolute grounding for a universal science, and the language used by Husserl remained under the aura of his philosophical roots. The "transcendental ego," "transcendental subjectivity," his "science of experience" with its "descriptive psychology," "apodicticity"—all retain the flavor of the transcendental traditions of Modern philosophy.

But the methods he evolved in the process soon threatened to break the old language and the old concepts as well. It remains Husserl's fate to have begun a revolution which soon outstripped him, but which must remain dependent upon him for many of its essential insights.

To note the philosophical interest in the relationship of phenomenology and hermeneutics, a preliminary observation about Husserl's strategy must be noted. As I have already pointed out, Husserl's problematic arose and took shape within the already constituted language and terminology of transcendental philosophy with its "subject" and "object" and its problem of how knowledge is constituted. On the explicit level Husserl took up this language and addressed himself to the set problems of this tradition. Yet implicitly almost every struggle Husserl had with the tradition and, for that matter, with himself pointed in another direction. In fact, often the results of the struggle pointed in directions exactly opposite to those which had been held by the tradition itself.

Perhaps the best example of this strategy and the problems it created may be seen from Husserl's famous *Cartesian Meditations,* the mature development from a set of lectures Husserl delivered in Paris in an attempt to interest French thinkers in phenomenology. The strategy employed, partly apologetic, partly polemic, was to utilize the explicit model of Descartes's version of philosophical method, whose notion of "doubt," a reduction by analysis to "clear and distinct ideas," and the establishment of the "ego cogito" as a base served as both model and foil for Husserl's own phenomenology. On the one hand, phenomenology was portrayed as being "like" Cartesian philosophy with its own method of "suspension" as a modification of "doubt," its own analysis into clear and distinct "givens" of intuition as a modification of geometrical method, and its own version of the subject as "transcendental subjectivity." But on the other hand, in and through each of these similarities, phenomenology arrived at conclusions directly contrary to those found in Cartesian philosophy. What phenomenological "suspension" showed was *the ultimate indubitability of the World;* what the analysis showed was that *"givens" are in fact constituted* by a complex process and not simples; and what "subjectivity" revealed ultimately was the *intersubjectivity* of the transcendental.

In general, some version of this strategy is followed in all of Husserl's mature writings. He begins with what is—seemingly—familiar, accepting it in a provisional way, but also undercutting it by placing it in "brackets," as an object to be examined and taken apart from a new and different perspective. In this process, the layers of the—seemingly—given object are unlayered so that both the object and the process by which the object is constituted are discovered. Here is a latent archeology—a *hermeneutic* process—which deconstructs what was previously taken for granted.

But such a process contains an inherent flaw. By taking what is—seemingly—familiar, situated in an already developed language with its accompanying terminology, the choice is one which allows the nonneutrality of the already developed language to retain its strength in spite of the attempted deconstruction through phenomenology. This unwanted and unexpected result plagued Husserl during his working life and remains extant in what to my mind are the still present misunderstandings of Husserl. Critics of his phenomenology still cry that this method is "subjectivistic," that it is a revived "idealism," that its domain extends no farther than "psychology," and the like, in spite of Husserl's intention to avoid precisely each of these epithets.

In this respect, Husserl must be termed a *naive* hermeneut. He did develop a powerful archeology of meaning in and through his "phenomenological reductions," but he failed to avoid, or transcend, or cut through the universe of discourse which retained its power through and in spite of his attempts to overcome it, precisely because he was not able to neutralize the nonneutral themes which pervaded the language of Modern philosophy.

Husserl's hermeneutically oriented followers were quick to discover this weakness and promptly attempted to overcome this strategy. Martin Heidegger, at least as early as *Being and Time,* was keenly aware of Husserl's strategic weakness and while adapting what I have argued elsewhere is a quite explicit phenomenological method for his own fundamental ontology also sharply diverged from Husserl's linguistic naïveté. Viewed in one way, Heidegger's strategy was to avoid as thoroughly as possible the extant problems of Modern philosophy by coining a radically new language which skirted or circumvented the terminology of "subject"–"object" and the constitution of knowledge. Heidegger "existentialized" each of the Husserlian steps, still following the Husserlian method quite clearly, but substituted a new sense at each stage. "Intentionality," which was Husserl's term for the essential correlation between the subject and his world, became "Being-in-the-World"; Husserl's "ego," transcendental or otherwise, became "Dasein" with the specific etymologically literal sense of being-here (being-there). The relationship between Dasein and World was no longer solely or even primarily a *knowledge* relationship, but a matter of existential dimensions, "existentiales." I shall return to these modifications in more detail in a moment, but here I would like to underline the essentially different and more self-aware linguistic strategy employed by Heidegger: the radical hermeneut.

This radical Heideggerian strategy also reverberates in an interesting way with the Ordinary Language strategies of the contemporary era. The "new" language which he coins is, in one sense, only new to philosophy. He sidesteps the usual terminology of Modern philosophy by turning to ancient or revived meanings or even literalizations inherent in ordinary language. Clearly Heidegger's language is not ordinary, but it is from the wealth of human history and language that he derives a richness lacking in Husserl and in much technically constructed philosophy.

Paul Ricoeur, another follower of Husserl, draws from a different strategy with respect to the transformation of transcendental phenomenology to hermeneutic phenomenology. He criticizes Heidegger's approach as too "direct." The new language world created by Heidegger, Ricoeur claims, too quickly cuts off the debate with classical issues (indeed, that is precisely what it was meant to do in one respect). Nor can one viably maintain Husserl's "idealism" or naïveté about the subject. Rather, one must create a *dialectic,* a debate in which two views may be interrogated, and a gradual approximation of terms occur such that the new meanings, a "third term," may arise from within the debate. Ricoeur's is a restorative hermeneutic.

There is a second and equally important issue in the movement from Husserl to his main hermeneutic followers (for our purposes here Heidegger and Ricoeur), which bears notice. There is a shift in what is taken to be of primary importance with respect to perception and language between Husserl and the hermeneuts. If Husserl borrowed from Modern rationalism its language of "subject," "ego," and the like, he borrowed the basically perceptualist notion of primordial evidence from the empiricists. Perception was the place where "primordial dator evidence" arises.

Both Heidegger and Ricoeur shift from this emphasis: Heidegger specifically maintains in *Being and Time* that perception becomes what it is only with respect to the fundamental contexts of language and discourse, and Ricoeur finds that all experience is already mediated *in media res,* in a world of symbols and myths and their interpretation.

The hermeneutic motto that "Man *is* language" expresses in part this shift from a perceptualist to a "linguistic" phenomenology. It also takes on its specific positive quality in contemporary hermeneutics.

However, in this context it is important to take some initial note of the specific contributions made by Husserl for the transformation of contemporary hermeneutics and its existential philosophy of language. These contributions may be made most clear, it seems to me, if one begins, not with Husserl's involuted strategies, but with the actual *results* of his phenomenology and with the key notions understood in terms of their *functions*.

The key notion in Husserl's phenomenology, around which everything else revolves, is *intentionality*. On the explicit (and therefore shallow) level of understanding, intentionality is simply the referential or directed nature of consciousness. All consciousness, Husserl says over and over, is consciousness of ______. Thus to perceive is to perceive something; to love is to love something; to imagine is to imagine something; and so on. This level of the interpretation of intentionality is simply that of a descriptive psychology of conscious processes. But neither is that the ultimate significance of intentionality nor does it reveal its true phenomenological function. At its depth, intentionality may be described as the *foundational correlational rule* of phenomenology by which any area of possible knowledge whatsoever is located and circumscribed.

In the deep sense of intentionality, the notion is neither psychological nor even strictly epistemological—it is ontological, the condition of the possibility of there being either "subject" or "object" or, for that matter, "World." Subject and object arise within what Husserl sometimes calls the *correlation-apriory* which began as intentionality.

In simplest terms, this means that intentionality as a primitive rule, a correlational rule, takes account of the interrelatedness and interdependence of what in the Modern tradition has been called "subject" and "object." Neither term makes sense in and of itself, and neither term can, in fact, be dealt with in isolation from the other. The whole of Husserlian philosophy revolves around this central correlation scheme, although the terminology and refinements upon how the correlation is to be understood vary from period to period. Take note of just two instances of how the ontological correlation occurred in Husserl's work.

First, there is the early correlation of what Husserl called "noema" and "noesis." Put most simply, "noema" is *that* which is experienced, the *what* of experience, the "object-correlate." "Noesis" is the way in which the what is experienced, the experiencing or act of experiencing, the "subject-correlate." Thus, there is no "noema" without "noesis" (there is nothing which is present *as evidenced* unless it is present to experience), and there is no "noesis" without a "noema" (no act of experience without that toward which it is directed.) One may formalize this relation thus:

noesis ⟶ noema.

Later Husserl was to complicate this bare scheme with his "egology" and make of the correlation a three-part correlation as in the *Cartesian Meditations,* where Husserl retained the sense of correlation but added to the act of experiencing a specific carrier of the act, the "ego." In this form the correlation took the form

ego–cogito–cogitatum.

In the basic transformation from Cartesian philosophy, this form of the correlation was one which maintained the essential insight of the earlier version of the correlation: no activity of experience (cogito as active, the thinking of the ego) without something which is thought (the cogitatum). The "addition" is the ego itself as that which does the thinking. But in essence, with or without an ego, the correlation remains functionally the same.

Husserl himself did not fully apprehend the results of using such a correlation analysis—at least as those who followed him did, for the implications of the correlation for the traditions of Modern philosophy are quite radical. These may be pointed up most drastically by contrasting them with the taken-for-granted conclusions of the Cartesian tradition.

First, on the side of the "object-correlate," for phenomenology there can be no such thing as a "worldless" subject. Doubt may be cast upon how to interpret the world, that is, the question remains open as to whether the world is basically "material" or "mental" or whatever may be the case in the metaphysical tradition, but *that* there is a world as that which is present constantly to experience cannot be doubted. A constant presence of something that is "there" exists, whether it is the construction of a demon or the creation of a god or the eternally given accidental matter of the universe. The world sense is primordially the sense of phenomenological presence and is *as indubitable as the ego.*

Second, on the side of the "subject-correlate," there is a negative result. There is no "subject" without a world, but neither is there any immediately self-transparent subject. The subject, within the phenomenological correlation, is deprived of its singular immediacy and of its presumed self-evidence. Positively put, *the subject now can know itself only by means of the world.* This implication of the correlation was, I believe, never fully grasped by Husserl himself, but it was grasped quickly by both Heidegger and Ricoeur, not in contrast to the trajectory of phenomenology, but in keeping with that trajectory.

Husserl already knew, from his investigations, that phenomenologically one must *begin* with the world, the noema, that which appeared, and from that work back *reflexively* to the subject, the noesis, the act of experience which is indicated by and through the world. That is, the shape of the experience is indicated by the way in which that which is experienced shows itself. Thus in the primitive correlation an order of procedure is established such that the noematic analysis must precede the noetic analysis:

Noesis noema
(2) ⟶ (1)
←---

The noesis, the act of experiencing, is not only "second" in the order of analysis—it is known only *reflexively.*

But because Husserl had adapted the already traditional metaphysical framework of "layers" in which there was to be an ultimate ground, he presumed that that which was arrived at ultimately in terms of the *order* of the analysis must be that which was fundamental. This was the source of his so-called "idealism." If we use the second form of the correlation, the founding stratum becomes the ego itself:

⟶
ego-cogito-cogitatum
(3) ←--- (2) ←--- (1)

Here the order proceeds from the cogitatum, reflexively (backwards) toward the ego which Husserl took to be the ultimate, founding stratum. He did not note so clearly the implicit limitation and darkness which his correlation and its reflexivity placed upon the ego.

Heidegger and Ricoeur *did* recognize this implication, and each in his own way turned phenomenology toward hermeneutics. First, let us take very brief note of the way in which Heidegger adapted almost entirely the Husserlian functional model of analysis in *Being and Time.*

Functionally, the Heideggerian analytic of Dasein operates exactly like Husserl's primary correlation. In this case the overall correlation looks like this:

(Dasein) being-in-the-World.

Recall that the preliminary task of *Being and Time* is to analyze the being of one being, Dasein, which is the kind of being we ourselves are. But to accomplish this task, Heidegger like Husserl and unlike Cartesian philosophy and its accompanying tradition, turns first to the World. Indeed, what must come first for Heidegger is a clarification of the "worldhood of the world." Dasein is to be uncovered in terms of its world. In short, Dasein is not to be known "directly," but "indirectly," by means of its world. Here we have the same form as the Husserlian analysis:

(Dasein) being-in-the-World.
(3) ←—— (2) ←—— (1)

But although the correlation and the order of analysis is isomorphic with Husserl's analysis of intentionality, there are two major modifications in the Heideggerian version of the correlation.

First, every element in the analysis has been "existentialized," that is, interpreted as components and structures of existence, rather than "merely" of psychology and epistemology. The formal ontology of Husserl has been replaced with a fundamental or "material" ontology in Heidegger. Second, Dasein, that to be analyzed, is not just the last term in the order but must be uncovered through the process of the *interpretation* of the phenomena, reflexively, but now a reflection specifically termed hermeneutic. That is why in this case I have placed parentheses around Dasein. Dasein has become enigmatic. And as the world of Dasein changes, so does Dasein.

In addition to the two modifications from the Husserlian base, one other addition specifically relevant to the hermeneutic direction given to phenomenology by Heidegger may be noted at this point. "Being-in as such," Heidegger's term for the correlation itself, is broken down into three equiprimordial dimensions: state of mind, understanding, and discourse. These are the primordial ways in which Dasein is related to World. Note that with the primacy of *understanding* and *discourse* one might already say that this primary relationship is one which is already *language*. Language is being-in, or language, existentially, is what constitutes the peculiar and particular character of our being-here. To analyze Dasein is to analyze language in its existential sense.

Thus for Heidegger one can say "Man is language," or, to invert the idea, one can note that language is a way, a mode of being in the world.

Paul Ricoeur, in a more indirect way, affirms the same direction toward a hermeneutic philosophy. In contrast to Heidegger, Ricoeur retains the language of the subject and remains in the center of the debates about how self-knowledge is constituted. His strategy is to develop a dialectic between phenomenology and some other, usually "objectivist," strategy. Each of the nonphenomenological strategies is seen to *limit* the pretensions of Husserlian naïveté, on the methodological level, and the pretensions of the ego, on the ontological level. For Ricoeur it is not only the case that the ego is known only reflexively; there is also a temptation to self-deception and narcissism on the part of the subject to avoid authentic self-knowledge.

Of particular interest are the recent analyses of Ricoeur found in his *Freud and Philosophy* and *The Conflict of Interpretations,* in which Ricoeur interrogates a whole series of what he calls "hermeneutics of suspicion," interpretations of the self which dethrone and cast into doubt the claims of both Cartesian self-transparency and any direct form of self-knowledge. Ricoeur sees what amounts to the whole "Hegelian left" and subsequently influenced methods as constituting a common strategy of attempts to pierce through "false consciousness." Nietzsche, Marx, and primarily Freud each find ways to break through the pretensions of the ego and find what is presumably unexpected, underlying various surface manifestations. Moreover, each employs a basically "linguistic" strategy in which there is some underlying "different" meaning to be brought out of some innocuous or even seemingly contrary surface manifestation. To cite one familiar example: Freudian analysis, as laid out in *The Psychopathology of Daily Life,* sees neither accidents nor neutral significance in jokes and slips of the tongue, but expressions of deeply held "unconscious" attitudes and beliefs. Ricoeur, in his hermeneuticizing of Freud, sees in this theory not an accurate metaphysics of the self (the machinery of super-ego, ego, id), but a sophisticated linguistic strategy which elicits the "grammar of desire." The implication for phenomenology, at least in its naive Husserlian sense, is that *more is meant than is intended in each expression,* and thus a hermeneutic process is needed to explicate the unsaid.

But in a deeper sense, Ricoeur has developed precisely one of the latent implications of Husserl's reflexive correlation. The subject does not and cannot know itself simply in Husserl's sense either. It knows itself in terms of its world, its "Other." In the psychoanalytic situation a specific and dramatic instance of such reflexivity occurs. The patient slowly and painfully learns *through* an other, the analyst. It is the analyst who brings out and allows to be made present for self-discovery the hidden meanings and surplus meanings which the patient him- or herself and by him- or herself did not and could not discover. Ultimately, for all phenomenology, subjectivity is intersubjectivity, but it is never that *immediately.* The radical decentering which occurs in the various "hermeneutics of suspicion" point to the difficulty of this process. But the process itself is essentially hermeneutic, an ongoing interpretation.

Ricoeur has taken this idea specifically into the notion of a *text.* In his more recent works, *La Métaphor vive,* for example, what reemerges is once again an affirmation of what was latent for hermeneutics in the

Husserlian correlation. What is it to understand a text? I believe I can point to one illustration which simultaneously shows how Ricoeur both reaffirms the phenomenological correlation and relates to a standard problem of interpretation.

The so-called intentional fallacy, which holds the impossible ideal for understanding a text by means of understanding the author's intentions, would be and is repudiated by Ricoeur in an essentially phenomenological hermeneutic. Just as the "subject" is not known either first or by itself in Husserlian phenomenology, neither can it be so known in textual analysis. Rather, what is presented is the "world" of the text. That is the matter to be understood first. To be sure, a "world" may show—indirectly—something of the author, even something of the author which the author may not have known directly about him- or herself, but whatever is known or can be found out occurs here reflexively, by means of the "world" of the text.

For Ricoeur, as for Heidegger, "Man is language." But language is enigmatic, often equivocal, and always multidimensioned. This is so for the subject as well, and in this language is "like" the self.

In this brief and at best schematic glimpse at the impact of phenomenology upon hermeneutics, I have concentrated on only one central set of problems, the central notion of intentionality and its transformation into the ontology of being-in-the-world, with its related problems of world and self as they unfold within a hermeneutic investigation.

I have suggested that a founding dimension of being-in-the-world is seen by hermeneutic phenomenologists as discourse or language. Thus language is neither something merely "inside us" nor "outside" us, but rather an *existentiale.* This is why Heidegger can speak of language as the "house of being" and of "language speaking" *(Sprache spricht).*

But before turning to the last task in this essay, let me note the connection between the hermeneutic philosophers and the earlier problems and traditions I left in the nineteenth century. The "general hermeneutics" of Schliermacher, with the primary task of developing an account of historical understanding overall, and the *Wissenschaften* problem of understanding, in Dilthey as a foundation for the social-historical sciences, are in effect continued by Heidegger and Ricoeur. Phenomenology has provided a rigorous methodology, neither "subjectivistic" nor "objectivistic," by which to approach social, historical, and human phenomena. Moreover, in its hermeneutic guise it takes as its focal dimension language in its deepest existential significance. The prospect for phenomenological hermeneutics lies in its own development of the philosophy of language.

3. Prospects for Hermeneutic Phenomenology

The philosophical focus of phenomenological hermeneutics, then, is an existential philosophy of language. In turn, an existentiale of language returns to the most ancient hermeneutic tradition. Hermes and the Christ signify a movement from the wordless to word, and the hermeneutic of language has turned to the problem of the "birth of meaning."

Poetry, as Heidegger conceives it, is primordial word. It is a "saying" which allows Being to be seen. *Poiesis* is creative-uncovering word which lets Being "shine forth." And poetic word, that which is most ancient, is united with that which is most radically futuristic. Such a word is "horizontal" in the phenomenological sense. This is to say that the coming into being of meaning becomes the focus for a hermeneutic philosophy of language.

Ricoeur has exactly the same vector in his thought. His long hegira toward the "poetics of the will" through the sciences of language center on the analysis of equivocity and ultimately *polysemy*. Poetic word, most dramatically apparent in symbols and myths, is the richness of polysemic language. But language itself in any word has this possibility. It is the productive imagination which enhances this polysemy, and it is to the imagination that Ricoeur has turned today. Hope, Ricoeur's leading religious idea, takes its shape through the imagination of the possible in poesis. Thus Ricoeur's philosophy of language, too, centers upon the poetic with a restorative and projective trajectory.

Both Heidegger and Ricoeur retain the thought of a coming-into-meaning symbolized by the ancients and revived by the Romantics in the motto of Johann Georg Hamann, "Poetry is the mother tongue of language."

Ours is, however, a very unromantic era, and the question arises: is a search for a word from Hermes or a recognition of Word made flesh possible in the late twentieth century? The gods, it would seem, have long since lapsed into silence, and even the notion of a sacred text has fallen into the relativity of a plurality of "religious utterances." Does "language speak" for contemporary hermeneutics?

I do not intend these vast questions to lead to a long meditation, but I do wish to suggest a few quite concrete problems which maintain the restorative-projective trajectory of hermeneutic philosophy in the present.

I should like to suggest not that a "first word," the "birth of meaning," "poiesis" fixes something which is exclusively "then" or "yet to come," but that its secrets are mundane and lie among us. Restoratively, one task of an existential philosophy of language is to recollect for us an understanding of the ways in which human discourse, speech, is *embodied*, the way meaning becomes flesh.[2] Projectively, this same task is one which must take into account the genuinely new embodiments of language which encompass us today. I think this includes, as only Heidegger in this tradition saw clearly, the question of technology as a hermeneutic problem. On the surface it may seem strange to link these two lines of thought, but underneath it seems to me that hermeneutic phenomenology demands that they belong together.

Our era, as philosophers have intuitively known, must necessarily come to grips with language precisely because language in contemporary times has exploded in a double sense. Language has become both a saturation of "languages"—not only native tongues, but also new tongues—and it has become newly incarnated in ways never known to humans in previous times.

It is true that ordinary discourse in a native tongue remains the coin of everyday life. But in a megasocietal and megatechnological world we all learn multiple "tribal languages" of the techniques which our various scientific and academic "tribes" bring into being. Nor is this explosion limited to ordinary and technical "languages"; it has burgeoned into a whole series of "artificial" or constructed languages such as those of mathematics, the symbolic logics, and various computer languages. And even if we do not "speak" or "read" such languages, we must live under their aura. Additionally, the way speech now becomes embodied no longer is restricted to that which is face-to-face or to listening and direct voice. Instead, we have amplified and extended our listening and voice through telephones, radios, and the current popular CB radio. Writing, the ancient nonverbal language of texts which posed an ancient hermeneutic problem, is now adumbrated with the equally transformed spatial-temporal reproductions of cinema, television, and tape recorder. As of yet there has been no full hermeneutic investigation of this series of embodiments. These new "texts" call for new types of hermeneutics.

This explosion of language into "languages" with its accompanying diffusion into technological embodiments saturates and changes our "World." And if phenomenological ontology is correct, when one correlate (the World) changes, it implies a change in the other (ourselves), but this change can only be understood by understanding the change in the "World."

I therefore suggest that contemporary hermeneutics in its trajectory of restoration-projection may both focus upon the recollective moment which recalls the primitive sense of incarnate speech and anticipate the futural sense of extended speech.

In the restorative moment, incarnate speech is focally and nominatively that which is face-to-face and is primarily embodied in heard and spoken word.

Richard Palmer in his book on *Hermeneutics* points out this simple but primary phenomenon which gets overlooked:

> interpretation is a form of saying. Likewise, *oral* saying or singing is an interpretation. In Greek times, *hermeneia* could refer to an oral recitation, of Homer for instance. . . . [T]hrough his intonations [he] "interprets" him, conveying more than he realizes or understands. He thus becomes, like Hermes, a vehicle for Homer's message.[3]

And in relation to the biblical root of hermeneutics, "the Bible is not information; it is a message, a 'proclamation' and was meant to be read aloud, meant to be heard" (19). Thus, Palmer concludes: "[Hermeneutics] has led us back to the primordial form and function of language as living sound filled with the power of meaningful utterance. Language, as it emerges from nonbeing, is not signs, but sound" (20).

What Palmer points to is the primitive embodiment of language. Speech is sounded, and sound is neither neutral nor reducible to a single layer of determined meaning. The "birth of meaning" lies in the *speech event,* and its normative embodiment is as a phenomenon of silence, listening, and sound. There is always more than information in spoken word.

Yet the very formulation of the contemporary sciences of language skew our observations in a different direction. Rather than a phenomenological emphasis upon word made flesh, the linguistic sciences divide the speech act into the two regions of meaningless sound (acoustics and related sciences) and into formal disembodied semiologies (semiology, semantics, grammar, and the like). Moreover, by taking such a turn, the temptation is to model language more and more upon the notion of "information."

A phenomenological hermeneutic in its restorative moment must, among other things, restore the sense of the normative embodiment of language in the concrete discourse of listening and voice. It must rediscover the nonneutral and rich significance of "musical" word.

But the dialogue of face to face, while primitive and normative, is clearly not the only manifestation of language for contemporary humans. Increasingly, language is extended in technological mediations. Television, cinema, telephone, and a whole array of language and language-related instrumentation transform the texture of our ordinary lifeworld. Time and space take on different significations as "objective" or "geographical," and even "temporal" distances become collapsed into the near distance of instant or repeated communication. As Heidegger rightly claimed, technology is in fact a kind of being. It can better be understood in the sense of a language-analogue than as a thing-analogue. Here, too, I see a task for a hermeneutically oriented philosophy, one which must deal with the projective possibilities of a technological civilization in which language no longer is discourse on the model of face to face. Such a projection cannot remain satisfied with the paradigm of immediate discourse but must revert to the other side of hermeneutics, the notion of the text through which past worlds become known to us.

Only now interpretation of a text across past temporal distance cannot remain the only direction for contemporary hermeneutics. It must also turn to the "possible worlds" of the future. Such an exploration in a radical sense, of the imaginative hopes and possibilities of humankind—and particularly those becoming horizontal in technological society—is called for as a prospect for hermeneutics. I am calling for interpretation not only across the past, but also across the future, in which one concrete and necessary task is the "science fiction" of a possible hermeneutic. In short, the projective hermeneutics is one which looks at "texts" across possible futures, the futures made available in technological culture.

The double movement, already begun in Heidegger and Ricoeur, may continue in such prospects, for of hermeneutics there is no end.

2

Language and Two Phenomenologies

Introduction

I have three concurrent concerns in this chapter. The first is to display a picturable model of some of the main features of phenomenological method. I wish in this case to clarify some of the complexities and implications of a phenomenological procedure for a philosophical context often more Anglophilic and Europophobic than not. But on the way to this end I wish also to begin the sketch of what I hope will become a considered reinterpretation of phenomenological history. I wish to differentiate two distinguishable, but often confused, lines of development from a common base in Husserlian thought. These types of phenomenology may be called, respectively, existential phenomenology and hermeneutic phenomenology and are initiated in their essential forms, respectively, by Maurice Merleau-Ponty and Martin Heidegger. Third, I wish to indicate briefly how the question of language lies imbedded in these distinctions. My outline of phenomenological development, then, wishes to account for the recent and increasing interest in the philosophy of language currently displayed among phenomenologists.

Because the tradition of European-born philosophies of consciousness are often given to excesses of self-consciousness, I shall remain true to form and bring to public awareness some underlying polemic concerns as well. Permit me three gross generalizations. First, existentialism, as it has been understood particularly on this continent, in a quasi-literary and often romantic guise is all but dead philosophically. This was the common opinion among the French professors I talked to in recent years, and although death throes may last longer just as birth pains start later in the United States, the symptoms are beginning to appear here, too. Herzog in the novel of the same name (and still placing Heidegger in the existential category) proclaims that "We must get beyond Heidegger." Second, this is not to say that existentialism in more philosophical form is dead. Quite to the contrary, as existential *phenomenology* it is just now beginning to make its appearance. As Sartre and Camus recede, Merleau-Ponty emerges. The previous guilt by association with the cultically "very existential" is replaced by concerns with more philosophically traditional problems revolving around perception, the problem of the body, and language. Last, despite the difficulty in dissociating the popular meaning of existentialism from its philosophical basis and the even more difficult task of removing Heidegger from that category, I would point out that today most Europeans are beginning to argue that Heidegger is actually and not just polemically correct when he disclaims ever having been an existentialist.

To counter these historical misconceptions and to construct a reinterpretation which apologetically helps to point up the properly philosophical dimensions of phenomenology, I wish to propose a new tripartite framework of understanding. Again permit me three general points: (1) Since all phenomenology in its most precise and recent formulation forms a constellation around Edmund Husserl in spite of the counter-gravitational pulls of Hegel, Kant, and Descartes, not to speak of Hume, I propose that we begin to think of phenomenology as a movement from Husserl to various forms of neo-Husserlianism. (2) To date the main lines of two distinguishable directions of neo-Husserlian phenomenology are *existential* and *hermeneutic* phenomenologies. Merleau-Ponty most clearly exemplifies the first, and Heidegger I now wish to cast into the second type. (3) Both of these developments have sources in the central and original phenomenological model developed by Husserl and particularly from the two sides of his notion of intentionality as the main structure of consciousness. Paul Ricoeur, one of the foremost Husserl interpreters, gives us the key from which I extend my interpretation. He says:

> In Husserl's first works . . . consciousness is defined not by perception, that is to say by its very presence to things, but rather by its distance and its absence. This distance and absence are the power of signifying, of meaning. . . . Thus consciousness is doubly intentional, in the first instance by virtue of being a signification and in the second instance by virtue of being an intuitive fulfilling. In short, in the first works, consciousness is at once speech and perception."[1]

In my interpretative extension of this observation I wish to show that existential phenomenology draws its strength from what I shall call the implicit perceptualism of Husserl's concept of intentionality.

Thus, although using radically different interpretations, existential phenomenology has motivations not unlike those of empiricism.

In contrast, I wish to show that hermeneutic phenomenology draws its concern from problems of language, also inherent in Husserl's concept of intentionality. This development is most clearly evident in Heidegger's work. It is a use of phenomenology which turns to problems of history and culture and which tends to read experience and ultimately perception itself in terms of a tradition of interpretation.

To outline this interpretation I turn first to a description of a general phenomenological model as a type of reflective philosophy. This type of model has its roots in the Husserlian origins of phenomenology, and as a general model is the origin upon which variations are worked.

1. Phenomenological Models: We Do It with Mirrors . . .

Reflective philosophies with parentage in the line of Descartes, Kant, and Hegel before Husserl usually are described as philosophies of consciousness or sometimes as subjectivisms. But often what is involved with reflection is not clarified as a methodological notion. Reflection in our language may mean some contemplative or rethinking activity—or it may mean more literally a reflection from a mirror or another reflective surface. My entry into a description of phenomenological reflection lies in a deliberate choice of what I would like to call *the metaphor of the mirror.* When asked how I do phenomenology, I reply, I do it with mirrors, because, as we shall see, in both thinking as reflection and the reflection from the mirror, what is arrived at is done so *indirectly.*

In fact, the mirror metaphor may first help clarify a persistent confusion among many who first read phenomenological descriptions as if they were simply revivals of introspective psychology. The key to the difference lies not in the material dealt with, since phenomenology deals with both so-called introspective contents and extrospective contents, but with the use of reflective indirectness.

To make this model yet more precise, I wish to introduce an analogy between the "I" and "eye." In relation to the history of epistemology with its play upon the subject and the object, phenomenology is a method which strictly reduces or restricts itself to a relational or bipolar understanding of the subject and the object. Husserl's *ego-cogito-cogitatum,* Heidegger's *In-der-Welt-sein,* and Merleau-Ponty's *Être-au-monde* are all versions of this essential characteristic of phenomenological method.

Thus a rule for phenomenology is that there is no subject without a corresponding object, nor is there a knowable object except for and related to a subject. This insistence upon maintaining a relational scheme is actually the functional heart of Husserl's so-called transcendental idealism. But its theoretical function serves as a normative concept which differentiates phenomenology from both the classical realist and idealist frameworks. The *epoche* or phenomenological reduction, understood in this light, is the means of strictly maintaining the bipolarity and "bracketing" the world in relation to speculative metaphysics. The aim is to suspend presuppositions, not experience.

This suspension of belief is on one side antirealist in the sense that phenomenology brackets the idea of an existent which exists apart from experience. But functionally this suspension merely establishes all contents of experience as "objects," or better, object-correlates *for* experience in order that they be displayed for description prior to fitting them into an explanatory or constructive schema. All cogitata or contents of experience become noema or object-correlates and are to be considered only in their relations to a subject. This means that the implied theory of evidence which takes shape in Husserl's philosophy is one which weights all immediate or fulfillable experiences as prior to constructions. In this sense phenomenology is potentially open from its very beginning to an "empiricist" direction, but an empiricism which remains distinct from its British and American relations in several ways.

But if realist assumptions are bracketed by the strict reduction to relationality, the same goes for classical idealism. The ego or the I nowhere exists in phenomenology without a world or field of contents which are the object correlates. Thus, Husserl's supposed idealism must again be seen as a methodological function rather than the assertion of the primacy of mind. Unlike, or short of, Descartes, Husserl's suspension of the world is not an active doubt, but a suspension of presuppositions concerning how the world is given out to be. The world is always to be described as it gives itself out to be to the knowing subject.

In both these senses the first central feature of the phenomenologically reduced model is the strict maintenance of a relationality between the I and the world. Once this reduction is presupposed, the metaphor of the mirror with its analogy may be reintroduced as follows:

> *The eye is to the mirror*
> (or reflective surface)
> *as the I is to the World.*

And if we were to ask the question How does the eye see itself, or the

I understand itself? the answer will be, Only reflectively or indirectly. It sees itself as reflected in the mirror, and I understand myself through the world.

This question, in its Husserlian context, establishes the *order* of procedure. Husserl's descriptions begin with the noematic or object-correlates of the relation. Thus, in the metaphor of the mirror what is first noted is that which appears immediately or naively "out there" as the appearance in the mirror. The naive level of awareness is always first directed to what is "out there" or in the world. This choice is one which makes a certain degree of sense in relation to both the history of civilization and the history of the individual. The prescientific aborigine and the preacculturated child may not be aware of our usual inner-outer distinctions, but they are aware of a certain "objectivity" to appearances. Thus dreams and spirits as well as rocks and trees are "objective" in the sense that they are "out there" as they appear to the individual. Phenomenology takes this pretheoretical naïveté to its extreme in the description of a layer of experience later called the *lifeworld* in Husserl's writings. Thus under the restrictions of the reduction the first focus is upon the contents of experience as they give themselves out to be primitively. What appears in the mirror is first to be described.

But in the case of the metaphor of the mirror a complication is entered. The face directly before the mirror is mine, and the eye reflects my eye. A turn may be made to the subject as the experiencer—but made by means of the reflection. Again, the history of the individual and the civilization display in rough outline such a turn. The child at some point learns to recognize himself in the mirror; the reflection is seen as a reflection of himself in a level of awareness perhaps never attained by the parakeet. It is from the reflection, the bouncing off from the object-correlate, that the eye is to be seen, and it is from the experience of the world that the I is to be understood.

It is precisely this reflective turn which differentiates Husserl at one blow from Hume, whom he admired so strongly. Husserl's critique of Hume's introspective psychology which finds itself unable to find a subject is located in the lack of a reflective turn. The ego does see itself-reflectively. From this phenomenological indirectness Husserl attempts to establish the structures of experiencing and upon the experienced.

It is well known that the structure of *intentionality* became the key concept around which the understanding of experience was elaborated in the Husserlian sense. Revolving around this central concept were a cluster of other notions such as "horizons," the "ray" of attention, and passive and active syntheses. But again it is the relationality of object and subject polarity which establishes this structure. Here we may put together both moments of the metaphor of the mirror.

If I first look at the mirror and observe carefully, I can note that the whole of the appearing surface is one which does not ordinarily appear with a flat or equal value. Usually somewhere near the center of the mirror the phenomena "stand out," whereas those on the fringe are less explicit. The eye spots the eye looking back. And if I were able to remain completely restricted to a naive and probably hypothetical level of awareness, I would end up saying that the center of objects in my field of vision are "more real" or "more important" than those at the edge. But even in an ordinary context the reflective turn is already made, and I say instead that my *attention* is what is focused. It is I who "make stand out" what I will under my gaze. Permit me here to make several leaps and conclude at the risk of prematurely losing specifics that this notion of a ray of attention within a wider or surrounding field becomes a picturable model for the general structure of experience in the Husserlian context. Experience is not only selective in the normal case, but displays itself as directional or referential.

But more important in the present context is the fact that the structure of experiencing is established by means of the reflective turn, and all experiencing is to be read via or upon the world of appearances. There is no subjectivity, phenomenologically speaking, apart from a world. But what is initially taken as a reflective surface makes a great deal of difference.

2. Multiple "Worlds" or the Hall of Mirrors

From a description of a general model based upon an interpretation of Husserl, I move to its variants with the neo-Husserlians. Husserl himself evolved a number of ways toward phenomenology, and he characterized himself as a beginner to the end of his days. At least three of these ways to phenomenology have become watermarks in traditional Husserl interpretations. The so-called early Husserl found his way to phenomenology by means of *eidetic* sciences, that is, mathematical and logical questions, and in this period he belongs to the set of philosophers who were concerned with the problems of logicism and psychologism at the turn of the century. The "middle Husserl," particularly of the *Cartesian Meditations,* tried for a while to describe phenomenology as a type of egology modeled in certain respects upon a Cartesian paradigm—but with some essential differences. It was during this period that the questions of solipsism were at their height. The "late Husserl" began to take a third major direction, increasingly perceptualistic in its form, in the concept of the *lifeworld*. It is here that he comes closest to what was to become an *existential* phenomenology.

I shall not enter into the various arguments in current Husserl scholarship which now often revolve around this enigmatic period, other than to make two suggestive indications about possible relations to the neo-Husserlians I shall examine momentarily. First, it is possible and perhaps even likely that Heidegger, already published and working out his own phenomenology, may well have influenced Husserl himself in relation to the *lifeworld* concept. Second, it is demonstrably the case that Merleau-Ponty concentrated his work upon the so-called late Husserl and took the lifeworld as a primary concept in the development of his own version of phenomenology.

Rather, I shall return to the structurally generalized model I have suggested and indicate how the neo-Husserlians began to vary this model to their own uses. The differentiation into an existential and a hermeneutic phenomenology depends upon which of the dual foci of consciousness is taken as primary, perception or signification, upon which focus serves as a "world." Thus, if I maintain my mirror metaphor, what is taken as the reflective surface is of great importance. I shall argue that although there are obvious overlaps between existential and hermeneutic phenomenologies, that Merleau-Ponty makes use of a "perceptual mirror" and Heidegger a "linguistic mirror." I shall, however, deal with these variants in an order inverse to their actual history since I believe that in some respects Merleau-Ponty comes closer to an extension of Husserl than does Heidegger, who couples phenomenology with a quite different set of problems than those motivating Husserl.

A. Existential phenomenology as a perceptualist philosophy

Merleau-Ponty began with the "late Husserl," who had already turned increasingly to problems of perceptual experience. It was a turn which, accelerated in Merleau-Ponty's thought, formed the possibility for existential phenomenology. Merleau-Ponty put it bluntly: "Far from being, as one might think, the formula for an idealist philosophy, the phenomenological reduction is that of an existentialist philosophy."[2] But in turn, an existential philosophy is one which focuses upon and makes more explicit the perceptualist side of Husserl's basic model. The distance to the world of the "early Husserl" becomes the presence to the world (*Être-au-monde*) of Merleau-Ponty.

But a perceptualism was implicit from the beginning in Husserl's theory of evidence. For one thing, the very use of language in Husserl drew its strength from perceptual metaphor. The phenomenological process was described as "viewing phenomena"; experience was focused as a "ray of attention"; things appeared in "perspective variations"; and even introspection was termed "inner perception." But more essentially, the whole reduction was one which as a theory of evidence increasingly weights immediacy or concrete experience as primary. Only that which is "bodily present" or which can be fulfilled in the intuition (experienced) shall be accepted as evident.

In Merleau-Ponty this direction becomes more apparent so that he makes the claim that "perception is primary." This phenomenological "empiricism," now known as existential phenomenology, is one which attempts to understand all human behaviors upon the basis of or in relation to phenomenology of perception. Of course, the theory of perception which Merleau-Ponty developed was one in keeping with the reflective model I have described. It is the use of this model which differentiates his "empiricism" from its Anglo-American cousins. He rejected all sense-data theories and systems of causation as constructions rather than descriptions of perception and turned to a description which was closer to gestalt psychology and perspectivism as the basis for his understanding of perception.

I would note, however, that Merleau-Ponty's existentialism *is* this version of perceptualism. His use of phenomenological reduction is one which attempts to reach a level of pretheoretical experience, the experience of the (perceptual) lifeworld. He says, "By these words, the primacy of perception, we mean that the experience of perception is our presence at the moment when things, truths, values are constituted for us; that perception is a nascent *logos;* that it teaches us, outside all dogmatisim, the true conditions of objectivity itself; that it summons us to the tasks of knowledge and action."[3]

This theory of evidence, regressive in its drive toward primitivity, remains in line with its Husserlian sense. To gain the perceptual lifeworld, the suspension or critique of all merely taken for granted interpretations or presuppositions is to be undertaken. In this process Merleau-Ponty's arguments against the empiricists attack the doctrines of sensation as merely constructs which falsify primitive experience rather than explain it.

But, as Merleau-Ponty notes, the reduction is never complete because what it arrives at is always the essential assumption of being already in a world, and it is this assumption which the phenomenologist wishes to explicate. The regressive direction of Merleau-Ponty's use of phenomenology nevertheless retains the bipolar model which remains essential to its Husserlian understanding.

To this point, Merleau-Ponty may be seen as essentially phenomenological in his use of a basic model. What differentiates him from Husserl is in what he takes as the world and what it reflects as a subject. His "mirror" and his "eye" are distinguishable. The world, for Merleau-Ponty, is the perceived world. It is presumed to be concrete, expressive, and rich in its contents. It is also a "natural" world rather than the strictly bracketed world of Husserl's formal ontologies. In this sense Merleau-Ponty tends toward a realism more strongly than did Husserl.

The same pattern is repeated when the reflective turn is made. If the perceptual lifeworld is the object-correlate for Merleau-Ponty, one would expect a symmetry in relation to the subject who must be thought of as a concrete perceiver. And this is the case. The problem of the body takes on a central role for the reflected side of existential phenomenology. The *embodied* or incarnate subject is the perceiving counterpart to the perceived world. *Le corps vécu,* usually translated as the lived body, is the perceiving subject in a perceptual world, and the concrete finitude of the body corresponds to the perceived presence of the world. Merleau-Ponty retains the bipolar Husserlian model but interprets it through perception.

One could argue, however, that this extension of Husserl is legitimate and that existential phenomenology is but carrying on the program of the "late Husserl." But that is not quite the case. By the selection of a "perceptual mirror" Merleau-Ponty chose only a partial aspect of Husserlian intentionality as basic. The perceived world and the perceiving embodied subject selects as primary the experiential side of intentionality as its basis and tends toward making signification secondary.

This becomes more apparent when one examines Merleau-Ponty's view of language. Even if his theory of language never attained the degree of development that his theory of perception did, it remains the case that it embodies a direction which is necessitated by his perceptualist base. It is also the case that as he turned more and more to the problems of language that this perceptualism was called into question.[4]

At first the Pontean theory of language seems to parallel clearly his theory of perception. Just as perception begins with gestalts and unitary wholes, so the world is seen as primitively expressive. And just as the perceived world is always understood only in its relation to a subject, so is language considered only in relation to the extant, speaking subject. In contrast to attempts to deal with language as if it were an empirical object, Merleau-Ponty insists that one begin with phenomenological speech.

But as this theory of language develops, a second meaning to Merleau-Ponty's phenomenological empiricism begins to emerge. His use of phenomenological reductions was one which always attempted to move back toward the perceptual lifeworld which at base becomes unspeakable. In return, speech becomes the counter-movement from the "silence" of the perceptual world to its expression in language. Thus Merleau-Ponty argues that language arises out of silence, out of gesture, and even that it is metaphorically a kind of music. This theory of expression parallels the aim of a phenomenology of perception to a pretheoretical world. The theory of expression is one which seeks a *prelinguistic* basis for expression. Thus, the whole weight of the Pontean theory of language concentrates upon the birth or coming-into-being of expression, the movement from silence to speech.

I shall not try to show here how this theory of languages follows from certain aspects of phenomenological investigation nor make complaint about some of the problems which I believe Merleau-Ponty raises. But I would point out that it is because he takes the perceptualist base as primary that this theory of language follows.

B. Hermeneutic phenomenology as linguistic philosophy

Here I abruptly leave Merleau-Ponty, suspended as it were, and turn to the second main neo-Husserlian direction as it takes shape in the thought of Martin Heidegger. My interpretation parallels that which I used with Merleau-Ponty and continues the metaphor of the mirror. The mirror Heidegger uses is the world of language in a broad conceptual and historical sense. And the side of intentionality which is developed is that which begins with problems of signification and meaning. Heidegger's presumptuous fundamental *ontology,* to my way of thinking, is a phenomenology of language, and it is from this phenomenology of language that Heidegger ultimately reads experience itself. This reading of Heidegger which sees him as hermeneut emphasizes aspects overlooked in interpretations which would make Heidegger an "existentialist."[5]

I begin by pointing up the frequent use of linguistic-historical metaphors in Heidegger's writing. For Heidegger, man is discourse *(Rede);* he is called *(Ruf)* to listen to the voice of Being; the famed destruction of the history of ontology is to be by means of a violent examination of philosophical language; and so on. In his backward or regressive aim at phenomena, hermeneutics or interpretation becomes the vehicle to be used. Heidegger explicitly claims, "Our investigation itself will show that the meaning of phenomenological description as a method lies in *interpretation.* . . . The phenomenology of Dasein is a *hermeneutic* in the primordial signification of this word, where it designates this business of interpreting."[6]

The essentially linguistic or conceptual strategy is revealed from the very beginning of *Being and Time.* Instead of turning to Husserl or announcing in strictly Husserlian fashion a program and a method, Heidegger formulates his phenomenology by his own version and translation of Greek terms. The phenomenon becomes "to let that which shows itself to be seen from itself" (the manifest). And logos, again translated in Heidegger's sense, is *discourse* or language in a primary and broad sense.

But for Heidegger phenomena are mostly covered over, hidden, and the task of phenomenology is to let these phenomena appear. On closer reading, it becomes apparent that what covers over the phenomena is *the history of interpretation,* the way in which language is used. Thus, to recover one's hearing of the voice of being one must loosen up and rework this covering. It is here that the long task of the "destruction of the history of ontology" begins and takes the long and varied detours via Kant, Descartes, Aristotle, and the pre-Socratics. Hermeneutics has as task the reopening of language to its fundamental or "original" possibilities. For Heidegger the categories of thought we employ have become hardened and prevent us from seeing, or better, from hearing, what underlies our categories as basic phenomena.

The whole context of interpretation which appears therapeutic in some respects is one which, through the distinctions of authenticity and inauthenticity, discourse and idle talk, seeks to uncover phenomena by hermeneutics. Heidegger wants us to "get back to" discourse, but to do so we must recognize and break through idle talk or mere abstraction. But what is important in this outline is to recognize that the way this is done is to take the element of language for the surrounding element which constitutes the world. It is from this recognition that we may understand the mirror Heidegger uses.

The functional outcome of choosing a linguistic mirror has more than one result. But it also changes the context of phenomenological questions. The substitution of a language world for a perceptual world moves what forms worldhood from the natural world to the cultural world. The field of silent objects is replaced by the field of human expression, the world of subjects. Second, the language world is one which relocates the focus upon intersubjectivity from gesture to the fullness of language. Intersubjectivity here is the necessary given from which one begins, for there is no private language to be found.

Third, one may expect to find, as Heidegger has already so well emphasized, that I am used by language as much as I am able to use language. But this is recognizable in the phenomenological context most precisely if one turns first to the language world as it gives itself out to be. One may expect to find that in certain respects, just as the concrete positionality of an embodied subject is constituted by the perceptual world, so will the hermeneutic subject be constituted by his language world, and the task becomes one of specifying in what ways and to what extent this occurs. Second, one may expect to find that the subject changes and has a history in relation to the language world. We are already aware that subjectivity is differently understood today than it was at the birth of our culture and that the "self" of a child is quite different from that of the adult. There is a history to the subject just as there is a history to the culture. In fact, one may expect to find with this mirror historical considerations playing ever stronger roles as the language world is noted in its subtle changes.

But, finally, one may expect to find in a counter-fashion something about the way language (or thought, if you will) changes what is perceived. The world may be potentially expressive, but what it expresses is different for us than for others in certain respects. The reflective model applied to language has not yet been exhausted. All of this lies within the scope of Heideggerian hermeneutics. For my purposes here, however, it is sufficient merely to point up the differences in problems and directions implicit in the difference between existential and hermeneutic phenomenologies.

3. Mirrors and Ontologies

If the neo-Husserlians seem to take different "mirrors" from which the subject is reflected, and if, as both seem to claim, what is arrived at is an ontology of human existence, it would seem we are back to a hall of mirrors. Although a series of mirrors may, by their very selections and exaggerations, reveal possibilities previously unexpected, each remains partial and short of finality. It might seem that phenomenologists now must choose between existential and hermeneutic directions, between what is taken as basic: experience or language? immediacy or history? And it might seem at worst that this is a return to sheer metaphysics with resurrected, if modified, empiricists and idealists again dividing up.

I would suggest that this is not quite the case, and I return to a comment by Paul Ricoeur, whose notions began this chapter. Ricoeur argues that phenomenology of speech and a phenomenology of language are two quite different matters and focus upon quite different aspects of the whole linguistic phenomenon. It is precisely this difference of task and focus of attention which holds the key to at least some of the differences between Merleau-Ponty and Heidegger.

By beginning with the "silence" of perceptual experience and by viewing language as expression, as the coming-into-being of significance, Merleau-Ponty in effect begins a phenomenology of speech. The subject struggling with language, to say the new, to express himself, is the focus of the perceptualist's immediacy. But by beginning with what has been said and by showing how we are used by our language and our interpretations, Heidegger begins with a phenomenology of language.

The phenomenology of speech and the phenomenology of language belong together. Immediacy without history is silent; history without speech is empty noise. And if the issues meet in a question of language, however differently that question is formulated on the Continent than in Anglo-American circles, it is not mere historical accident. The "linguistic turn" now belongs to the hedgehogs as well as to the foxes, and the philosophy of language animates Paris today just as it does Oxford. And phenomenology through or in spite of its excursion through existentialism is returning full circle to Husserl. But this is a response to our time. I close with a comment by Ricoeur, whose earlier comment suggested this line of interpretation:

> Today we seek a broad philosophy of language which would be able to account for the multiple functions of human signification and its mutual relations. Now is language capable of such diverse usages as mathematics and myth, physics and art? It is not by chance that we ask this question today. We are precisely the ones who use symbolic logic, exegetical science, anthropology and psychoanalysis, and who for perhaps the first time are capable of embracing as a unique question that of the reconstitution of human discourse. In effect the progress of the disciplines as disseparate as those mentioned have at the same time manifested and aggravated the dislocation of this discourse. The unity of human speech is a problem today.[7]

3

Philosophy of Technology as Hermeneutic Task

By combining philosophy of technology with hermeneutics, I am pairing a newcomer with an ancient discipline. As Paul Ricoeur has so often reminded us, hermeneutics goes back at least to Aristotle's *Peri hermeneuia* (On Interpretation). This classical notion became narrower and more technical in premodern European history insofar as it became a particular theory of interpretation of *texts,* more particularly, *biblical texts.*

With modernity, however, there is *hermeneutic expansion,* which although still related to theological concerns, began to become coextensive with one version of *humanities or the human sciences.* In the cases of both Friedrich Schleiermacher and Wilhelm Dilthey, what had been textual interpretation was transformed into a "philosophical anthropology," with the former, and into the sciences of *Verstehen* with the latter. In both cases, however, the humanities or human sciences with principles of "understanding" or interpretation were differentiated from the natural sciences and principles of "explanation."

This expansion continued into the mid–twentieth century with a new trajectory borrowed from *phenomenology.* Indeed, the three primary philosophical hermeneuts—Martin Heidegger, Hans Georg Gadamer, and Paul Ricoeur—all fused hermeneutics with some version of phenomenology (hencefore referred to as P-H traditions). Heidegger's version was central insofar as he took *ontology* itself to be essentially hermeneutical. And while his version of a hermeneutic ontology, both linguistic and historical in *Being and Time,* could be called "direct," Gadamer's and Ricoeur's versions were, in contrast, more "indirect." Nevertheless, all three principals relied upon the now combined P-H tradition of hermeneutics.

Such a P-H tradition remained within the aura of the humanities, or human sciences focus. And even if the structure or ontology of the human was claimed (Heidegger's *Dasein*), phenomenological hermeneutics took as its subject matter such phenomena as language, history, the socius, philosophical anthropology, and—in a kind of revival of an ancient theme—*texts,* albeit texts of all types, including the metaphor of the text, which began to define the late-twentieth-century offshoots such as poststructuralist and deconstructionist work as found in Michel Foucault and Jacques Derrida.

The welding of *phenomenology* to hermeneutics, however, also included a focus not previously found in premodern hermeneutics, and that was the emphasis arising from both Edmund Husserl and Maurice Merleau-Ponty on *perception, praxis,* and *embodiment.* I shall contend in this chapter that this set of emphases opens hermeneutics to a new trajectory, one which now takes hermeneutics into yet another expansion in the late twentieth century, namely, the realm of *science and technology.*

That this is a new expansion and direction is evidenced through an interesting set of confrontations which occurred just last year in Budapest. One of the interesting results of the collapse of Marxism as the "official" philosophy of the Eastern European countries is the sudden release of interest, particularly among the younger philosophers, in a revival and modification of the P-H tradition. Thus, a group of younger P-H philosophers in 1993 organized a first international meeting on hermeneutics and science in Hungary. Some thirty of us attended this meeting, with larger contingents from Hungary, Germany, and the United States, but with others as well.

As it turned out, two quite different strands of approach began to take shape precisely between the issue of hermeneutic expansion I have mentioned. The two principals on the one side were Karl Otto Apel and Dagfinn Føllesdal, who argued for what I shall call the "older" form of P-H hermeneutics, and on the other side my colleague Robert Crease and myself, along with several of the younger European thinkers including Lazslo Ropolyi, one of the organizers, arguing for the "new" expansion.

The Apel-Føllesdall stand insisted that all hermeneutics was essentially human science oriented—thus one could have a *hermeneutic history of science,* but one could not have a *hermeneutics of science or scientific objects* as such. In short, one could do history of science, sociology of science, anthropology of science, but not anything like a hermeneutics either of "nature" or, for that matter in this context, of "artifacts" or technologies.

The Crease-Ihde-Ropolyi stand argued that there could and was now a *hermeneutic* philosophy of science, which included hermeneutic analysis of scientific products and of technological artifacts. Thus hermeneutics was to be expanded to the nonhuman, the inorganic, and the artifactual as well, but in a unique P-H sense. The arguments were sometimes quite fierce, with Apel characterizing the expansionists as "those Americans"!

Historically, of course, a minoritarian strand of a hermeneutic philosophy of science and technology has arisen. It includes such philosophers as Joseph Rouse, Patrick Heelan, Joseph Kockelmans, Hubert Dreyfus, Robert Crease, and myself (this name list does not exhaust P-H philosophers by any means but points out some who have published books with this thrust.)

I have undertaken this brief history to bring hermeneutics to the frontier with the philosophy of science and technology. But now I shall do the same with the philosophy of technology in particular, the newcomer in the pairing of the theme.

Philosophy of technology, even with a somewhat longer history in Europe than in North America, is a relative newcomer to philosophical subdisciplines. In Europe, the usual earliest citation is that of Ernst Kapp, a neo-Hegelian, who published *Grundlinien Einer Philosophie der Technik* in 1877. One can also refer to Karl Marx's interests in modes of production in the same period.

In a recent book, *Thinking through Technology*,[1] Carl Mitcham traces out two strands of the philosophy of technology—interestingly not too far from the divisions noted in the older and newer P-H traditions. The one is what he calls an "engineering" version of philosophy of technology, the other the "humanities" approach. In this case, 1927 turns out to be a significant year. It was the year of the publication of both Friedrich Dessauer's *Philosophie der Technik* and Martin Heidegger's *Sein und Zeit* (Being and Time). Dessauer, an engineer and a neo-Kantian, exemplifies the first and Heidegger, the phenomenological ontologist, the second. One should note in passing that although Dessauer's work is thoroughly focused upon technology as a theme, *Being and Time* is not. Rather, *Being and Time* lays the groundwork for what only subsequently becomes the full-blown philosophy of technology in Heidegger's later development. The tool analysis in the discussion of *Vorhandenheit* and *Zuhandenheit*, based upon the primacy of praxis, does set the context for the later developments in "Die Frage nach dem Technik" and other later works.

Both the engineering and humanistic traditions continued in Europe, and, indeed, technology and technologization became something of a focus between the world wars and continuing after the Second World War. Virtually every major European philosopher—Heidegger, Jaspers, Berdyaev, Ortega, the later Adorno, Habermas and the Frankfurt School, et al.—had something to say about technology. Much of this, however, tended toward the distinctly *dystopian* and saw technologization as a threat to older cultural and historical values.

If we now turn to the twentieth century, I wish to note just one more historical phenomenon: the traditions motivating the philosophy of technology, as contrasted to those motivating philosophy of science, were often quite divergent. Philosophy of science, both in Europe and in North America, underwent a period during which it was dominated first by positivism and later neopositivism, and even later by distinctly analytic traditions. These traditions were largely linguistic or propositionally oriented and placed theory building at the center of their concerns. Contrarily, and only much later as an institutionalized development, philosophy of technology remained linked, both in Europe and North America, to Marxian, neo-Marxian, P-H, and, in the American case, pragmatist traditions. These traditions hold in common a more *praxis* orientation and often elevate experiment to the foreground. Thus, there has been both a divergence of philosophical traditions associated with the two subdisciplines, and, in the case of contemporary institutionalization of philosophical societies, a time divergence (in North America the Philosophy of Science Association was founded in 1934, whereas the Society for Philosophy and Technology only officially organized in 1983).

1. Technology and Hermeneutics

The brief histories are in place, and philosophy of technology is at least juxtaposed with P-H traditions in philosophy. In what follows, I hope to show how a phenomenologically enriched hermeneutics bears upon technologies, and how this essentially hermeneutic process leads us into a deeper understanding of technological phenomena.

The focus which unites these two phenomena was one I mentioned above: it arises from the *phenomenological* emphasis upon *perception, praxis,* and *embodiment.* These emphases are to be found in all the "classical" phenomenologists. For example, when it comes to knowledge, phenomenologists argue that all knowledge is relative to embodiment. We can know only what is possible for a bodied being—and this implies an actional, perspectival, and situated kind of knowledge.

I shall not retrace the variant ways in which this position is worked out in Husserl, Merleau-Ponty, Heidegger, or Ricoeur, but I will note one aspect: with the exception of Heidegger, none of the classical phenomenologists pushed far the way an embodied being has a transformed sense of experience *through artifacts and technologies.* I note a few exceptions: Husserl did take brief account of how language changed with the introduction of writing; Merleau-Ponty did recognize with some sophistication the extension of bodily experience through artifacts—such as the blind man's cane and the feather in the woman's hat in *Phenomenology of Perception*—but above all, Heidegger recognized and described the way tool use is taken into a kind of prepredicative experience in *Vorhandenheit* situations.

It is by extending these insights, and by focusing much more particularly upon the human-artifact interface, that a genuinely P-H approach to philosophy of technology can begin. My own itinerary has followed this aim. Already in the early seventies, I had become interested in the ways *instrumentation* transformed knowledge gathering and production in the sciences. It makes a great difference if one can view the heavenly bodies only with one's naked eye or with the device of the telescope, and that startling difference was loudly proclaimed by Galileo in his "Heavenly Messenger" which publicized "what no man had seen before," and which became—as Derek de Solla Price called it—"an artificial revelation." In passing, it is interesting to note that Husserl's Galileo *is not a telescope user*, but a mathematizer! His concern was that a mathematizing science was one which gradually "forgot" primary perception and the body and that such a mathematized science was one which moved away from this lifeworld-inserted body into ever more abstract realms of ideality. I note in passing that this emphasis upon mathematization also belongs to the dominant or standard view of science within both positivism and any theory-oriented interpretation of science. This is a pre-Kuhn philosophy of science.

Yet the telescope-using Galileo was very much an *embodied* observer, albeit an observer who had a kind of "extended, although transformed" vision—a vision by means of and through the telescope. Only a science which forgets to take account of its own *embodiment in instrumentation* can "forget" the lifeworld.

It was to this, and related phenomena, that I turned in the early seventies. And, in a four-chapter sequence, I collected this work and published in 1979 *Technics and Praxis: A Philosophy of Technology*. This was and is cited as the first English-language philosophy of technology book published in a major philosophy of science series, The Boston Studies in the Philosophy of Science.[2]

At that time I was, one could say, "lonely." Today I am no longer lonely, since the role of instrumentation (and experiment) has become a major industry for some philosophers of science. I myself have dealt with a spectrum of both Euro-American and Anglo-American "instrumental realists" in *Instrumental Realism: The Interface between Philosophy of Technology and Philosophy of Science*. There, I added in addition to Heelan, Dreyfus, Crease, Rouse, and myself, Robert Ackermann, Ian Hacking, and

Peter Galison from the Anglo-American traditions, and, in an extremely provocative way, Bruno Latour.[3]

2. Hermeneutic Expansion through Technologies

Now to return to the themes which motivate this chapter. The question now becomes one of how a "new" hermeneutics can be expanded through a focus upon *technologies*. At the outset it can be seen that phenomenological epistemology is "materialist" at least insofar as it centers in actional, perceiving embodiment. And, on the other side, technologies are also material—at least as one component in the larger technological context. The "hardware" aspect of technology is its "bodily" characteristic.

It is at this very point, in the analogization of human embodiment with artifactual embodiment, that an *expanded hermeneutics* is called for. This is, in part, because we are still in some sense "moderns." Thus a preliminary problem might seem to be the question of how the two "embodiments" differ. For example, this is the debate which for so long raged between Hubert Dreyfus and the AI computernaughts.

AI computernaughts tended to be modernist physicalists; one could hope for an "artificial intelligence" because both brains and computers are physical processes, and the big job was to make the electronic device simulate the "meat" or "wetware" of the brain. I shall not rehearse all of Dreyfus's arguments about why computers could not do this, but the clearly P-H conclusion Dreyfus drew was that the computer could not be intelligent because *it did not have a body*, which is to say, a human body. Its materiality neither perceives, moves, nor acts. In one sense, of course, the computer does have a body, so much of the argument could also hinge on bodily differences between electronic and fleshly bodies. But this set of arguments remained within the modernist context. In spite of these limitations, the Dreyfus/AI computernaught arguments did take place in part in a P-H application to artifacts, to technoscientifically produced objects.[4]

Dreyfus's early work on this issue took place in the sixties (*Alchemy and Artificial Intelligence* was published in 1967). At the same time Patrick Heelan, a physicist-philosopher, began to argue that scientists perceived-read the world through their instruments.[5] And in spite of the fact that in his account such a perceiving-reading was taken to be virtually totally transparent, this was a different perspective regarding artifacts. In this case, artifacts were taken into and extended the "bodily" experience of the observer. Heelan was to later hermeneuticize this whole process in his *Space Perception and the Philosophy of Science.*

In one sense, one could say that what Heelan did was to have taken Heidegger's tool-use analysis and apply it to scientific instrumentation. This, at least, has the advantage of allowing not only simple workshop tools to be *Vorhanden,* but also complex instruments which in another context perform essentially in the same way. But, in Heelan's use, one difference also occurs: the "perceptions" which occur through scientific instrumentation are *measuring* perceptions. Perception and measurement (mathematization) come together in a scientific-hermeneutic process. Any Heelanean Galileo would thus have to be both a mathematizer and perceiver (observer). In his later work, *Space Perception,* for example, to utilize a "readable technology" such as a temperature thermometer is simultaneously to "read-perceive" and to "measure" the temperature.

Both Dreyfus and Heelan, then, have used P-H traditions to expand the hermeneutic into the nonhuman object, but in different ways. Dreyfus leaves the computer as "other," or better, simply always less than human. Its "thinking," its "intelligence" remains alien and nonhuman. Heelan, on the other hand, comes close to "humanizing" the artifact. Insofar as the instrument enters into an extended human embodiment, it becomes virtually transparent, so much so that Heelan claims that reading a thermometer is "equivalent" to a direct perception.

I shall now briefly turn to my own role in this context. As indicated, my work goes back roughly to the same period as those of my colleagues. My application of P-H traditions initially borrowed from all three of the classical phenomenologists: Husserl, Heidegger, and Merleau-Ponty.

From Husserl and Merleau-Ponty, I used an "intentionality model" of human-technology relations. As a human being, intentionality is that directedness and referredness toward an environment or "World." And, existentially, that being-toward-a-World is not simply conscious, conceptual, or cognitive, but actional, perceptual, and embodied. However, in the human case, although one can and often does *directly* relate to environments in embodied ways, as Heidegger had already noted in his tool analysis, so very much of our relation to an environment is *mediated through the use of tools or artifacts.* Does this make a difference, and, if so, what kind of difference?

In the context I am tracing here, I now position myself between my two contemporary peers: Dreyfus and Heelan. To caricature the situation, Dreyfus, I believe, leaves the technology *outside.* For him, not only do computers "think" differently than humans, but they also do not "think" at all. The computer remains an alien presence which, only through philosophic illusion, becomes a semblance of the human. Heelan, on the other hand, collapses his "readable technologies" too much into unique measuring perceptions of a particularly trained, although technologically extended, human body. Heelan's technologies are taken *inside.*

For me, technologies are neither *outside* nor *inside.* Nor do they reveal themselves *directly,* but rather *indirectly* or reflexively, in a distinctively phenomenological-hermeneutical way. Concentrating first upon technologies as artifacts, while it is possible to *use* praxically any material entity "technologically," that is, to take up the artifact into some human-directed or referential praxis toward the world, as in the example of simply picking up a stone and throwing it, for the most part humans more often do at least some minimal *technical* modification of the artifact before putting it into use.

Ancient traditional societies, probably more than two and a half million years ago, took stones and chipped them, and fashioned them into spears, or hoes, or scrapers.[6] And while the materials themselves were all "natural"—that is, the material base itself was not changed—there were minimal remakings of found objects into more suitable ones, fitting them *Vorhanden.* Contemporarily, of course, the process might well entail actual manipulation of *atoms,* making them into artificial molecules as the dream of contemporary *nanotechnologists* now project. Our reach, both at the macro- and microlevel, has obviously drastically changed in our human history. These observations, however, are not P-H observations.

Rather, what I hold to be a P-H model for interpretation is a *distinctively relativistic* one. By relativistic, I mean in a sense analogous to contemporary physics, the perspective which always has to account for *what* is observed, but relative to the situation or position of the observer. The object does not simply reveal itself to me, particularly not objectivistically.

In such a relativistic framework, it is the interrelationality which is crucial. Thus a P-H analysis must look at the realm of human-technology *relations.* But in the spectrum of such relations, as Heidegger observed as early as *Being and Time,* our uses of artifacts are such that the artifacts themselves do not stand out, and, indeed, when they stand out they are no longer functioning in the world-related way in which we experience them.

The relativistic or *intentionality* model also can show this. I may, of course, undertake *direct* or face-to-face actions, fully perceptual, actional, bodily. A simple case is to pick a piece of fruit which I see to be ripe from a tree and then eat it. But, if the apple (shades of Adam and Eve) is too high to reach, I may pick up a stick and knock it down, and then eat it. This action, then, is *technologically mediated.*

Were I to simply formalize these two sets of actions, I could diagram the direct action incident as

I ----▸ apple,

but, in the technogically mediated action, I could diagram the incident as

I–stick ----▸ apple.

As simple as this distinction is, there is much to take note of:

- There are basic differences in the direct and mediated situations, and they are experientially differentiable.
- In the direct action, I-am-my-body and simply pick the apple. But in the mediated situation, the artifact *occupies mediational position between the apple and me.*
- In the mediated situation, my "reach" is extended or "magnified"—I can do more than I could do in my naked body position. But, simultaneously, at least during the actual use of the technology, my experience of the apple is "reduced." This latter point is often overlooked in favor of the former magnificational point—but, for example, I do not feel the fleshiness of the apple, nor tactilely sense as fully its state of ripeness, etc.
- The mediated situation, then, is one in which both *what* is experienced and *how* one experiences the object are changed. *Technologies transform our experience of the objects in the world non-neutrally.*

I have rehearsed here some of the beginning of what long ago I called a "phenomenology of human-technology relations." I shall not repeat that analysis or go into the various structural types of such relations which were discerned. Rather, I shall turn to the expansion of a P-H methodology to show how it hermeneutically entices the transformational powers of the various technologies.

What can be intimated are some of the structural aspects which such an analysis yields. First, such an analysis shows that all technologies are *non-neutral.* Technologies change and transform situations, however minimally. There is no such thing as a "mere use" of technologies. Second, there is also a structural aspect to such non-neutrality; I have sometimes called this the magnification-reduction structure of technology use. Technologies are selective and do not simply replicate nontechnological situations. This is what makes them both useful and interesting. But this selectivity also both amplifies some feature of what and how something is experienced and reduces some other aspect. Third, while both of these points may be considered structurally "universal" for all technologies, each different technology displays a different pattern of amplification-reduction.

These three points are part of the indirect or reflexive way in which a P-H analysis expands hermeneutics, but they can also be used heuristically as a guide for what to look for in particular analyses. One expects, once practiced in such analyses, to discover patterns of magnification and reduction—and thus one can project how to compensate for, overcome, or adumbrate aspects, for example, if one is building instrumentation. But here we also reach a complication.

I have also argued that *all* technologies are multistable. If a technology is what it is contextually, relativistically, then any given technology not only will have multiple purposes but also will fit into any number of indefinite contexts. A piece of bamboo cane can be used to weave a shade screen for a tropical window, or as a fishing pole for a youngster, or as an instrument of punishment for a vandal, but while this is a multiple set of possibilities for a simple artifact, the same holds for the most complex of our technologies. What had been "intended" as the most complex single instrument for contemporary physics—the supercollider—may end up *being* an extended mushroom farm, as one proposal suggests. There is an essential ambiguity within all technologies.

Nor do the complications end with an "intrinsic" essential ambiguity, because technologies are always *culturally embedded,* that is, any given techology will also be culturally relative as well. The history of technologies exemplifies this all too well—the Indian prayer wheel become Lowlands water pump is but one example. These are, moreover, not simple matters of different uses. Cultural embeddedness is a matter of a technology-in-a-context where it *is* what it is only contextually. This is why there is also no such thing as a simple technology transfer. There is only a culture-technology transfer, for all technologies have a network of "assignments" as Heidegger called them.

3. The Hermeneutic Process

I have traced a brief history of how the P-H tradition has expanded and today is taking account of the realm of science, and particularly technology. What emerges is not an objectivist set of descriptions about kinds of technologies, or some equal categorization of the subjective set of uses, but rather a reflexive and relativistic characterization of the structures of nonneutrality, multidimensionality, and multistability of human-technology relations, or, better, of human-technology-world relations which begin to show something of the complexity and indeterminability of technology in its contemporary form.

These studies have been characterized by some as "postmodern." But if Bruno Latour is right, there is a sense in which we have never been "modern" by virtue of *practicing* differently than we have categorized. In his book *We Have Never Been Modern,* Latour comes to a set of conclusions quite commensurate with those I have been suggesting from an expanded P-H tradition.[7]

Latour suggests that the Modern penchant to differentiate strongly between Nature and the social or Nature-Culture creates a situation in which much, indeed if not most, of what happens and gets developed is a *proliferation of hybrids.* Although I have termed it differently, what I have called human-technology relations are also "hybrids." Provocatively, Latour includes in his lists of hybrids such phenomena as the ozone hole. The ozone hole is simultaneously "natural" and "cultural" if our science is right in claiming that about a quarter of its formation is of "homogenic origin."

It may be time for us, by rethinking and reframing, finally to leave both the Modern and the older and narrower understanding of hermeneutics as belonging to the human sciences as distinct from the natural sciences.

4

Whole Earth Measurements

1. How Many Phenomenologists Does It Take to Detect a "Greenhouse Effect"?

Let us take a very commonplace, often discussed, and critical topic within our conversations regarding critical environmental issues: Are we detecting a "greenhouse effect," and related to this, is it exacerbated by "homogenic factors," that is, human actions? I suspect that most would be inclined to give a positive answer to both of these questions. But, if pushed philosophically, what would be the evidence, and how well grounded would it be for such affirmations?

Within scientific communities and associated scientifically informed circles, the answers have to be somewhat more ambiguous, particularly when rigorous questions concerning evidence are raised. Were scientific truth to be a matter of consensus—and some contemporary philosophers of science argue that scientific truth often turns out to be just that—then it is clear that there is beginning to be a kind of majoritarian consensus among many earth science practicioners that the temperature of the Earth, particularly of the oceans, is indeed rising and that this is a crucial indicator for a possible "greenhouse effect."

Most of these scientists admit that the mean oceanic temperature has risen globally in the last several decades. But this generalization depends, first, upon how accurate measurements may be, not just for samples, but for *the whole Earth*. Hot spots, for example, the now four-year-old hot spot near New Guinea which is part of the "El Niño" cycle, doesn't count by itself because it might be balanced by cold spots elsewhere. And the fact of the matter is that "whole Earth measurements" are still rare and primitive in the simple sense that we just don't have enough thermometers out. Second, even if we have enough thermometers, a simply *synchronic* whole Earth measurement over three decades is but a blip in the *diachronic* history of Ice Age cycles over the last tens of thousands of years. Third, even if we know that the Earth is now heating up and has an ever increasing ozone hole, and from this computer-model strange weather effects can be predicted, how much of this is due to homogenic factors, such as CFCs, CO_2 increases, hydrocarbon burning, and the like? Is it the case, as *Science* magazine claimed in 1990, that "24% of greenhouse encouraging gases are of homogenic origin"?

As I have described the current debate it does not sound on the surface to be very philosophical; instead, it seems empirical and thus within the domain of scientific discourse. However, in the way in which the story was cast, some deep epistemological issues lurk which relate to my ironic heading, "How many phenomenologists does it take to detect a greenhouse effect?"

What I wish to do here is to look both at "classical" phenomenology (Husserl in particular) and at Heideggerian hermeneutics regarding our theme, "environmental phenomenology," and show that both approaches are to be found wanting with respect to the "greenhouse effect" phenomenon. Then I wish to show how a rather radically modified hermeneutic and phenomenological epistemology can address this problem. This modification entails two positive concepts, both of which, I will argue, are necessary for a phenomenological approach to environmental issues: *technoscience* as a thoroughly technologically embodied science as a necessary concept to get at presumed subperceptual entities, and *Earth-as-planet* as a necessary presupposition for dealing with whole earth measurements.

Classical phenomenology, I argue, lacks the first of these concepts, and Heideggerian romanticism rejects the latter. Let us begin with a very simple phenomenological question: from what standpoint or perspective can the issue of whole Earth measurements be made? If, at base, our very knowledge is constituted by way of our *bodies* and through *perception* as both Husserl and Merleau-Ponty seem to contend, then what is claimed about whole earth measurements becomes problematical in two senses.

First, how do we get a sufficiently encompassing perspective to talk about a whole Earth? Second, given that the greenhouse-producing gases are, in fact, not *perceived* at least in Husserl's primordial dator sense or at the level of primary lifeworld perception, how do we constitute these?

One approach might be to recognize that the scientific discourse concerning greenhouse gases seems itself to be constructed in terms consistent with *modern* as opposed to phenomenological epistemology. Indeed, all the entities to be measured—ozone, CFCs, CO_2—are subperceptual in any direct perceptual sense. One cannot directly *sense* them. In early modern terms (Cartesian) these entities are "inferred." And, indeed, these constructs, even in Husserl's version of Galilean science, are thought to be distant from the primary perceptions of the lifeworld and are, instead, results of a *mathematization* process.

I cannot take the time here to walk through the whole of Husserl's version of mathematization and Galilean science, but the elements which are of importance for this argument revolve around the virtually exclusive interpretation of science through mathematization. At the heart of Husserl's version is a strong distinction between primary perceived *plena* which are directly sensorily perceived and mathematized objects (the pure shapes of geometrization) which are both *abstract* or idealized, and only *indirectly* available:

> [Science in] . . . all this *pure* mathematics has to do with bodies and the bodily world only through an abstraction, i.e., it has to do only with *abstract shapes* within space-time, and with these, furthermore, as purely "ideal" limit-shapes. *Concretely*, however, the actual and possible empirical shapes are given, at first, in empirical sense-intuition, merely as "forms" of a "matter" of a sensible plenum; thus they are given together with what shows itself, with its own gradations, in the so-called "specific" sense qualities: color, sound, small, and the like.[1]

Husserl's mathematized science thus presupposes this strong distinction between the fully perceived *plena* and the merely idealized abstractions of pure shapes.

Thus, to incorporate the perceived world into and under the scientific world, there must be an *indirect* process:

> Now with regard to the "indirect" mathematization of that aspect of the world which in itself has no mathematizable world-form: such mathematization is thinkable only in the sense that the specifically sensible qualities ("plena") that can be experienced in the intuited bodies are closely related in a quite peculiar and *regulated* way with the shapes that belong essentially to them. (35)

In short, some form of association between the idealized abstracta and the plena is needed, a relation which Husserl terms *indirect.*

If we now return to our environmental problem, the measured changes occuring with greenhouse gases, were Husserl right then the gases are themselves constituted through *mathematization* with its abstractive process and indirect or inferential geometrization. But there is something radically wrong with this approach because it implicitly leaves out the very *materiality* of the gases themselves and with it any possible "realism" which drives the earth sciences as well as any other late modern science.

The flaw, however, lies not with the description or the presumed "indirect" constitution of the gases; it rather lies with *Husserl's flawed concept of science.* And while Husserl is not alone—he in fact remains close in his interpretation to many other "Cartesian" or theory-weighted interpreters of science, including the positivists of his time and Descartes himself—it is his concept of science which has undergone the *abstraction.* Of course, science requires *measurement, quantification,* and the processes of *analysis* which occur in mathematization, but it equally requires a *material* relation with the "things themselves," and this occurs in actually embodied science. That embodiment is the *technological extension of primary perception* through *instrumentation.* Husserl's Galileo is a Galileo without the telescope! And the relation between Galileo's satellites of Jupiter and rings of Saturn and early modern science is not "mathematical" or "indirect," rather it is *mediated* and *instrumentally real* through this technological extension of bodily perception. This technologically *mediated* perceivability reduces and transforms the presumed strong distinction between Husserl's plenary perception and the equally presumed indirect and abstract reduction to merely pure shapes. CFCs, CO_2, and ozone are not pure shapes but are instrumentally presentable, material entities through science's increasingly sophisticated technological embodiments. They, too, are plena, albeit micro-plena instrumentally presentable. Simply put, science's mathematization does not fit the Husserlian version of it, and, moreover, a purely or only mathematized science is itself a disembodied science. Contrarily, a technologically embodied science is a science which may operate through an *instrumental realism* which allows a *mediated* bodily perception of micro-entities rather than of abstract, pure shapes. I am arguing here that only such a *technoscience* can allow for the realism needed to take a greenhouse effect seriously.

2. Only by Being Profound Can Heidegger Be Profoundly Wrong

Husserl has not often been invoked for discussion of environmental issues. But Heidegger has; indeed, he is often termed "green," and he has been invoked particularly by "deep ecologists." There are two surface reasons why this is the case. First, in the example here, the greenhouse effect, if it is the case that homogenically originated causes arise from escaped CFCs, hydrocarbon burning processes, and increased CO_2 accumulations, it is largely because of industrial processes or "Technology" with the capital "T" Heidegger gives to it. Technology is thus culpable in the environmental crisis. The second reason for the invocation of Heidegger is a "romantic" one and lies in his honorific use of the term "Earth." (Has anyone noted that the Heideggerian concept of Earth as groundedness actually fits a still geocentric universe?) Finally, I find, however, that there is a deeper reason why Heidegger is thought to be relevant to the environmental crisis, which lies in his much more "material" realism concerning the artifactual, which I contended was lacking in Husserl.

This "realism" was very early applied to the role of artifacts in his concept of the *Zuhanden.* Tools or instruments were ways in which human actions were *mediated* into the World. The hammer "withdraws" into the sense of bodily action which produces the project in its materiality. Much later, in the context of an emergent philosophy of technology, Heidegger included science itself as a kind of institutionalized *Zuhanden* process. One must also remember that the tools which Heidegger analyzes in *Being and Time* play a special role there—they are the indicators of the "Worldhood of the World." The covered railway platform "takes account of the weather," and Nature itself is shown via these artifacts. But, later, technology, now elevated to a metaphysical mode of seeing, makes science itself "technological": "It is said that modern technology is something incomparably different from all earlier technologies because it is based on modern physics. . . . Meanwhile we have come to understand more clearly that the reverse holds true as well: modern physics, as experimental, is dependent upon technical apparatus and upon progress in the building of apparatus" (*Question,* p. 30). Science is technologically embodied. This is the origin of its instrumental realism, and although Heidegger did not follow through on this concept, he may be said to have thereby corrected the Husserlian "abstract" view of science.

In Heidegger's view, however, science is doubly instrumental. If, in its actions, it is through its instruments or artifacts that it can act upon nature, it itself is an "instrument" of a certain metaphysical way of seeing, "Technology" with the capital T. It is the means by which Nature is challenged and taken as "standing reserve" or a "resource well" [*Bestand*]. In short, it is through the artifactual or material instrumentality of science that this can occur. By extension, then, there can be a homogenically related greenhouse effect, through the very challenging process.

Normatively, this is part of the danger found in Technology; by being "materially efficient" it can pose a threat to the preservation of *Earth,* another of Heidegger's central concepts. Technology, now become materially efficient through the artifactual, reveals nature as a resource well which can be acted upon. Or, to put it another way, nature in the *Modern* sense becomes what it is through the possiblities of material instrumental realism.

Now I want to turn to another, but subterraneanly close, look at Heidegger's "materially efficient" or "realist" uses of the artifactual. Tools are not the only artifacts; Greek temples are also artifactual. And, like earlier *Zuhanden* uses of tools, they *reveal a world:* "Standing there, the building rests on the rocky ground. This resting of the work draws up out of the rock the mystery of that rock's clumsy yet spontaneous support. Standing there, the building holds its ground against the storm raging above it and so first makes the storm manifest in its violence."[2] This artifact carries lots of freight—it reveals, or even "makes" the storm reveal itself as storm, the very sun its grace, the light its light, the invisible space of the air . . . tree, grass, eagle, bull, snake, cricket "come to appear as what they are."[3] The temple-work (artifact) "opens up a world and at the same time sets this world back again on earth, which itself only thus emerges as native ground" (42).

But while this Wagnerian, Nietzschean romantic tone rings much louder than the accounting for the weather of *Being and Time*'s *Zuhanden,* it plays a similar role. It is through the artifact (work) that the world is revealed and now settled upon the Earth. The temple is a kind of "nice" technology! "To be a work means to set up a world . . . [but this ultimately also] lets the earth be an earth" (42, 46). Heidegger then goes on to create a dialectic between "world" and "Earth." Other than to reiterate that the dialectic itself is revealed through the work or artifact, I shall now leave the commentary and turn to what I think to be profoundly wrong with this view.

The first error lies in the romantic amplification of the deeper insight from the *Zuhanden* period. If it is through the artifactual in its praxical context that a "world" is revealed, then the later Heideggerian selective romanticization of this in his ontological aesthetic is one which fails to *fully* reveal the world which gives rise to the artwork, but which instead in its romantic selectivity *covers over* precisely the full set of involvements which the artwork potentially reveals.

It was the historian J. Donald Hughes who first drew my attention to this problem. Although I doubt he knew of Heidegger's work, he opens his *Ecology in Ancient Civilizations* with the following remark:

> Those who look at the Parthenon, that incomparable symbol of the achievements of an ancient civilization, often do not see its wider setting. Behind the Acropolis, the bare dry mountains of Attica show their rocky bones against the blue Mediterranean sky, and the ruin of the finest

> temple built by the ancient Greeks is surrounded by the far vaster ruins of an environment which they desolated at the same time.[4]

As I read this passage a new "revealing" suggested itself—instead of storm, snake, and bull, I began to see rock surpassing forest, species-count changing or disappearing, and soil irreversibly eroding into sea around Heidegger's temple. But there is worse: were one to look closely, much more closely, at the temple itself, one would find in its very stone testimony to the Greek destruction of an environment. The first temples were made of wood—which became increasingly scarce as deforestation continued, not under chainsaws but under mere bronze axes, probably well crafted in a *techne* way—and the temple on the hill evidences this past: "The Greeks persisted in using the architectural forms that they were accustomed to from the days of wood. All the details of the entablature, with its cornices, friezes, . . . were copies in stone of wooden structural elements. Greek builders even imitated the pegs that held the ancestral wooden structure together by adding little stone knobs called 'drops.' "[5] In short, even the temple itself, read closely, evidences the process of earlier deforestation, and all of this before "modern machine technology."

The temple as *artwork,* at least in Heidegger's description of it, does not, will not, perhaps cannot reveal its wider and more negative context. The reason, I contend, is the choice of a romanticized version of such works. But, deeper, the artwork is simply the other side, the obverse of a *technology* or a technological artifact. Both are material; both are focal entities which reveal fields of involvements and references; and both thus "gather the fourfold" to use a Heideggerian overstatement.

There is, in the Heideggerian context, good artifactuality—the artwork—which reveals the romanticized interplay of the gods and mortals, earth and sky, and bad artifactuality—technologies—which rush us across rivers (the steel bridge), or dam the same river (hydroelectric dams), or reduces the Holocaust to modern agriculture (both of which treat humans as standing reserve.) While I argue that there is something profoundly wrong with such romantic reductionism, I now want to make one more post-Heidegger move before returning to the greenhouse effect.

Good artifactuality—the temple—reveals Earth as ground, that upon which a world can rest. Bad artifactuality—in this case I shall introduce the now familiar *global imaging* made possible by such technologies as Earth shots from the Moon—makes of the world a "world as picture": "Hence world picture, when understood essentially, does not mean a picture of the world, but the world conceived and grasped as picture."[6] This, presumably, is bad. What makes it bad is the "reduction" of the world to picture, whereas world (itself) cannot be so reduced.

Behind this critique, of course, lies much ancient and still modern garbage, which beginning with Plato's copy-epistemology and its transformation into modern representationalist-epistemology, needs weeding out. Heidegger's insight that world is revealed as picture is better than that, but it still misses the point. I wish to argue, contrarily, that to see by means of imaging technologies is not to "look at a picture," it is rather to "*look through the image.*" That, at least, is how imaging—including whole Earth measurements—is used in scientific praxis.

3. Whole Earth Measurements

I now return to the initial problem: how can one tell anything about a greenhouse effect, phenomenologically? I opened this venture with the assertion that two concepts are necessary as a frame for detecting a greenhouse effect: science as *technoscience* and a perspective, *Earth-as-planet,* for making whole Earth measurements.

The first is necessary if there is to be any referential "realism" regarding the phenomenon. Measurements must be "measurements" of ________," and what an instrumentally embodied science gives one is precisely the opening to micro- and macrophenomena through the mediated bodily perceptions made possible by instrumentation. Instruments give access to the phenomena, and that often in mediated perceptual form. Greenhouse gases are not "inferred," they are instrumentally "perceived." I argued that the Husserlian science of the *Crisis* lacked this conception of science, although the praxical realism of Heidegger's earlier work saw its possibilities.

The second conception needed is *perspectival* in the sense that one must have a sense of a "whole Earth" as that field which is measurable, and this is what I am calling *Earth-as-planet.* The idea of Earth as planet, as a finite sphere, is itself quite ancient: it is anticipated by Aristarchus, it is assumed by Copernicus, and its size has been measured along with these notions, both in Greek and early Modern times. But one could argue that it is not *seen* as planet until it is so embodied in the earth shots with which we are now all familiar. And it is here that I enter the last set of arguments with classical phenomenology, Husserlian and Heideggerian.

Put crassly, what might Husserl and Heidegger see when they look at a Moon shot of Earth-as-planet, or when they look at other imagings of environmental phenomena? If the answers are "pictures" or "images," then the look involved is both naive and cast in terms of modern epistemologies which remain caught with a passive theory of perception which sees only "surfaces" and "representations." And, if in the Heideggerian case, there is more penetration, but a penetration which yields the referent phenomenon as itself a "picture," then the seeing remains short of *scientific seeing* as evident in science praxis. Indeed, such a look would remain short of even the primary insights of a phenomenological epistemology claimed by both but restricted to everyday or prescientific lifeworld regions. Heidegger notes, "What we 'first' hear is never noises or complexes of sounds, but the creaking wagon, the motorcycle. We hear the column on the march, the north wind, the woodpecker tapping, the fire crackling." This auditory plenum "precedes" its abstract resolution into distinct "qualia," the gestalt "precedes" its separable parts: "It requires a very artificial and complicated frame of mind to 'hear' a 'pure noise.' "[7]

Similarly, I am arguing that it likewise takes an artificial and complicated frame of mind to see a "mere picture" or an "image"—this is an effect of a modern epistemology. Instead, in scientific seeing, a "seeing which understands" to again paraphrase Heidegger, gestalts are "seen through" in the processes of imaging and measuring. And no scientific seeing takes the image for the referent itself, nor confuses the world with its "picture." Rather, all "pictures" and "images" are themselves instrumental and are "seen" or "read through." I return now to the greenhouse effect.

4. Science Praxis as a Phenomenological Hermeneutic

Contemporary *technoscience* does its work by means of its technological embodiments, its instrumentation. Its range of instruments is wide and diverse. With respect to some of these referring to whole Earth measurements, attempting to discern a possible greenhouse effect, there are various sorts of recording "thermometers" such as buoys in the oceans which record both surface and depth temperatures. Similarly, thermometers are placed in the atmosphere (balloons, aircraft probes, etc.). The recorded results are then fed into computer models, mathematically analyzed, and reduced to graphs and trajectories to determine overall patterns. This process falls under what I have earlier called "hermeneutic relations" whereby the referent is mediated by digital, numerical, or graphed results which are nonisomorphic with the items themselves referred to.

Imaging processes, however, are of special interest because imaging mediates perceivable patterns which often do refer isomorphically, or partially isomorphically, with the intended referent. Thus satellite photography in geosatellites *depicts* Earth phenomena which in more visually gestalted forms show the state of the "whole Earth." I choose this special form of instrumentation here because it displays a much more *phenomenological* morphology in its very closeness to perceivability.

In the cases of straightforward isomorphism, in which color photography remains "representational" with true colors, one can detect what is happening on the planet. One can "see" the deforestation process in South America, and a recent satellite photo dramatically showed the smoke plume hundreds of miles out to sea from the forest fires on my own Long Island. And, with enough satellites, one can scan the whole planet. But, scientifically, such straightforward visual isomorphism is not enough.

What I call "instrumental phenomenological variations" show more: thus by turning to the sophisticated varieties of "false color" which varies from human vision, one may learn more. Infrared imaging shows much more dramatically the changing ratios of organic to inorganic surfaces, thereby showing where desertification is taking place. Thermal imaging shows temperature variations; magnetometry shows even that which is below the surface; and with each variation a richer and more complex set of phenomena emerge. I am suggesting in the strong sense that this deliberate variational praxis is, in science, very much akin to an embodied form of phenomenological variational praxis.

But it is also critically "hermeneutic." One must learn to "read" the interrelations. For example, the current ability to get closer to *synchronic* whole Earth measurements must also be related to *diachronic* readings of the past. These are made possible by the material "calendars" left by the Earth itself. Tree rings reveal past periods of drought and wet seasons, deep glacial ice shows past temperatures, and so on. I am suggesting here a "hermeneutics of things" not merely of languages and texts. (Indeed, I take the current penchant to make of everything a "text" to be analogous to the Heideggerian fear of taking the world as "picture.")

Finally, scientific praxis—even if it still sometimes self-interprets its knowledge in modern epistemological language—is no longer modern. It is, rather, *reconstructive.* Instrumental perceivability is not passive observation, it is actively constructive. To get the "truest" imaging, the result must be manipulated, "constructed." Here two techniques may be noted in passing: first, the use of contrastive techniques is known in all types of imaging processes. False color or some other contrastive technique is used to sharpen, enhance, and bring forth features which otherwise would lie undetected. Here manipulation *away from the passively real* is needed to bring out the phenomenon. Second, computerized enhancement also may sharpen and create more "true" or "hyperreal" imaging. I contend that this "technologically constructed" approach is beyond the limits of modern epistemology and is more akin to a "phenomenological constitution" which in its classical sense did not realize its scientific potential.

Imaging, as one example of the technological embodiment of science, is not some set of "pictures," it is the praxical, instrumental *Zuhanden* of that which reveals a world. Insofar as science is not merely some set of mathematized theory, but is embodied as a *technoscience* which materially relates to a world, and insofar as the Earth is a perceivable *planet,* and then, insofar as our measurements are critically reconstructed through instrumentally constitutive praxis, *then* phenomenologists can detect the possible greenhouse effect.

PART 2

CONTINENTALS

A phenomenological hermeneutics necessarily draws not only from texts and languages, but also from *perceptions,* and that is why Maurice Merleau-Ponty is a hermeneutic fulcrum for much of what transpires in this direction. His notion of speech and language, arising out of bodily activity, is of highest import. Paul Ricoeur, perhaps the most open-minded of contemporary "continentals," also plays a crucial role in hermeneutic transformations, and so identifying his place in this history is also to be noted.

But, again, I conclude this section with another angle upon the imaging process. Continentally inspired philosophers have, in the late twentieth century, "textualized" much, perhaps too much. But the questions of "reading" and "texts" can take a different set of vectors when seen under the developments of imaging.

5

Singing the World: Language and Perception

Phenomenology is a revolution in man's understanding of himself and his world. But the newness and radicality of this revolution is faced with a problem, the same problem which arises in the epiphany of any new phenomenon. What phenomenology has to say must be made understandable—but what it has to say is such that it cannot be said easily in a language already sedimented and accommodated to a perspective quite different from that taken by the revolutionary. What eventually may be said must first be "sung." One only gradually learns to hear what sounds forth from the "song."

Not long ago an illustrative event of like dimensions occurred when the "songs" of whales were recorded. The listener, in every case known to me, would first be taken aback by the strangeness, the mysterious, enchanting, uncanny quality of the songs. Fascinated and even awed by this new language from the sea, deep stirrings of feeling occurred. Then a second phase of appropriation would begin through associations and metaphors: "That's *like* a bull bellowing," "It's *like* electronic music," "2001," "Now I can see how the legends of the sirens began." Here the listener attempts to relate the uncanny to something which is already familiar—and that's as far as most go. With the mysterious partly domesticated one is satisfied. But a further thought is possible: this is, after all, neither siren, nor electronic music, nor bull—it is the humpback whale sounding forth, "singing the world" in his own way. It is for us to listen, to enter that strange song as best we can if we are to discern the contours of that world. For this, more than curiosity is called for. The whale song issues a call to whose resonances we cannot yet fully respond.

Language, "singing the world," is a philosophical problem. But that problem is more acute for the phenomenologist. The phenomenologist, faced with ordinary language filled with the sediments of a past history, must learn to sing in a new way. Merleau-Ponty was quite aware of the difficulties of both saying and hearing something new in his struggle to express himself. Of philosophies he said, "I begin to understand a philosophy by feeling my way into its existential manner, by reproducing the tone and accent of the philosopher. In fact, every language conveys its own teaching and carries its meaning into the listener's mind."[1]

I want to pose the question of language as the foreground focus in this essay on Merleau-Ponty because language is on, if not is *the,* Archimedian point from which other questions may be levered from their background dominance. Behind, under, eventually beyond spoken language lies the world of perceived, wild being which is Merleau-Ponty's more apparent focus. For him it is through the question of perception that the question of language and expression is reached—but it is through the question of language that the enigmas of perception may also be seen.

Nor are the questions of language and perception separable for Merleau-Ponty. Phenomenologically the world is already primitively given as meaningful in some sense. There is no pure datum, no raw qualia or pure sense from which to begin; rather, man begins immersed in a world already significant, already both "natural" and "cultural," and the phenomena of immersion are the first to be interrogated. Thus if I begin by reversing Merleau-Ponty's emphasis it is because that reversal is suggested by his work itself. Insofar as perception and expression remain paired it is possible to begin from either side of the pairing, noting, of course, that a reversal of order may also reveal aspects not apparent from the other side.

The initial practical problem—how does one make phenomenology understandable?—is at base more profoundly philosophical. It is too easy for the philosopher already assuming and inhabiting the "phenomenological attitude" to hold that the problem is not one of language at all. Nor is it, if all problems of language are those of particular propositions or of logics and grammars. The problem is one of a shift in stance. Once one learns to "see" as a phenomenologist, then what has been said by Husserl, Heidegger, and Merleau-Ponty appears neither obscure nor ambiguous. The problem is one of attaining a perspective, not one of uttering a correct formula. But this shift of perspective *is* a problem of language in a deeper sense. In elaborating the non-neutral, embodied theory of language in *Phenomenology of Perception,* Merleau-Ponty noted that language "presents or rather it is the subject's taking up of a position in the world of his meanings."[2]

In this preliminary sense phenomenology, too, is the taking up of a position in the world of philosophical meanings. It, too, is situated as a focus within Being. It, too, "sings the world" in its own style. And our task is to listen to that song.

I shall attempt first to develop some of the characteristic marks of Merleau-Ponty's notion of language. By situating his work in the context of a philosophical tradition and noting some features of the growth of a radical language, by outlining the distinctive features of an explicit theory of language, and then by noting the role language plays vis-à-vis perception, I hope to show the justification for this reversal of emphases.

In the preface to *Phenomenology of Perception,* Merleau-Ponty explicitly situates his work in the context of the phenomenological philosophy issued earlier from Husserl. Merleau-Ponty sees his own development of phenomenology as a nuanced divergence from certain aspects of Husserlianism. First, the perceptual world is primary, the base from which one must begin and the primitive field which must be thoroughly explored. Second, the examination of this field will yield certain essential ambiguities about man and his relations to his world which are revealed better by a focus upon the *genesis* of meaning than by attaining a description of stable essences. And, third, the genetic emphasis will result in the development of an *existential* as contrasted with a transcendental idealist philosophy.

In situating himself alongside Husserl, Merleau-Ponty also adapts and refines a strategy used earlier by Husserl. *Phenomenology of Perception* employs a polemic against both empiricist and rationalist traditions, against their mechanist and intellectualist outcomes. Posed positively against these traditions is the emergent existentialist philosophy developed by Merleau-Ponty. This positive position belongs to that class of contemporary anti-Cartesian philosophies which reject the dualism of mechanical, material extended substance and psychological, subjective mental substance. In contrast, Merleau-Ponty's existential position elaborates a unitary theory of embodied being.

But taking a polemic position is taking a position within an already constituted world of philosophical meanings. There is a price to be paid: not only must the existentialist position which emerges be drawn in contrast to dominant philosophies, it must also first address itself to its opponents in the very language of those opponents. If Cartesianism is to be rejected, what replaces it? If sense data are rejected, what is perception? The form of the questions resituates us in the midst of a linguistic problem. If there is to be a new framework and a new language to express the insights appropriate to an existential phenomenological position, how are those expressions to emerge from the non-neutral philosophical past?

Three degrees of increasingly radical language uses are discernible: (1) one seeks to use standard terms and gradually change their meanings, to adapt them to new usages; (2) one creates new terms (neologisms, compound words) and employs metaphors to infuse new meaning; (3) and one borrows words from other contexts not previously used by philosophers. (One might add, that the result (4) would be an extension in which the new language will be a new technical vocabulary as the new tradition itself begins to resediment after the stirring it initially caused.) Merleau-Ponty utilizes all three of these levels, but in varying stages.

1. The early tendency is to be more conservative. *The Structure of Behavior* attempts to infuse Gestalt terminology with a nascent phenomenology. Within the polemic structure of *Phenomenology of Perception* the standard terms are again reworked. Indeed, key philosophical terms are often dealt with in the manner of the first use of language, but filled with new meaning. *Perception* is a significant example. Perception is primary for Merleau-Ponty, but perception becomes both broader and more inclusive than its previous philosophical use. His defense of the thesis of the primacy of perception before the Société Française de Philosophie is an attempt to widen the very meaning of perception:

> By these words, "the primacy of perception," we mean that the experience of perception is our presence at the moment when things, truths, values are constituted for us; that perception is a nascent logos; that it teaches us, outside all dogmatism, the true conditions of objectivity itself; that it summons us to the tasks of knowledge and action.[3]

Here *perception* must carry a load seldom imposed upon it by philosophy since its equivalent in the pre-Socratics.

2. Similar adaptations and transformations happen with *subject, thinking, imagination, object.* But the essentially conservative strategy of importing new meanings to old terms is surpassed by the introduction of increasingly suggestive coined terms and metaphorical uses. Already begun by Husserl and Heidegger, Merleau-Ponty accepts and expands the vocabulary of newly coined terms, *life-world, being-in-the-world, intersubjectivity, lived body,* and to these adds a series of at least seemingly metaphorical uses of *silence, incarnation, gestural meaning, singing the world.*

The coined terms are quasi-technical from the beginning, despite their odd ring to the uninitiated. *Lived body* (*corps vécu*), for example, is meant both to contrast with the objectified sense of body used in the sciences and to refer to a primary, nonreduced sense of living being as embodied being. Similarly, *intersubjectivity* is meant to contrast with the Cartesian state of private selves and to suggest positively that we are already "outside ourselves in the world."

Even as technical terms much of their strength comes from evocations not found in the traditional and key terms of (1). This evocative sense is stronger yet in the metaphorical uses. The silence which precedes and which indirectly conveys meaning within speech, "singing the world" which precedes the philosophical norm of conceptualized meaning, pushes us further toward a radical and different philosophical vocabulary.

3. Although there is a continuum of radicality from metaphor to the importation of words from spheres not previously used by philosophers, to my mind the most radical and interesting use of language comes into greater prominence with Merleau-Ponty's later writing. *The Visible and the Invisible,* itself an evocative and suggestive title, brings us *flesh, chiasm, intertwining, perceptual faith,* and *wild Being.* This is "wild language" more akin to a literary genre than to much philosophy. Merleau-Ponty's struggle with language leads him beyond the strategy of transforming terms to the initiation of a radical discourse.

What we are seeing here can be stated quite simply, although the reality which is embodied in this development of language is far from simple. Simply, as the implications of the phenomenological revolution become clearer, the more radical the use of language becomes. The new wine threatens to burst the older philosophical wineskin. We are witnessing a birth of philosophical meaning. This is the case even if we must allow that the rent of the wineskin may eventuate in the dissolution of philosophy as it was previously known; that, at least, is one possibility. The emphasis within Merleau-Ponty's theory of language upon the genesis of meaning is thus both a result of the demands of the phenomenological turn *and* a reflection upon Merleau-Ponty's own philosophical experience.

The anti-Cartesian polemic of *Phenomenology of Perception* forms the background against which both the emergent theories of perception and expression take shape. The argument which rejects both empiricist and intellectualist interpretations of perception and speech is one which attacks any notion of dualistic "purity." Both traditional interpretations maintain a *pure* (psychological) mental being and a *pure* (mechanical) material being. In contrast, the existential theory, based upon a phenomenological interrogation of experience, claims that such purity is constructed, not found.

Rejected are all notions of pure data, sensory or conceptual. Accepted is an essential ambiguity of the perceptual object, an incompleteness, an openness to multiple possibilities, all of which Merleau-Ponty argues is true to the actual perceptual experience. Rejected are both the objective (mechanical) body and a transparent intelligence (mind) as immediate interpretations of experience. Accepted is the embodied subject whose every action is subject to an initial movement from the unformed to the formed, whose gesture precedes any later attained clarity of intellection. Rejected is an objectively given world whose reality is merely to be discovered (constructed) by the right method of formal geometricization. Accepted is a world which is always pregnant with significance, but whose meaning must be rewon through an interrogation of its presence. The lifeworld appears between the subject and the world within the focus of perception.

The outcome of this line of thought on a functional level is to recognize that any theory of the body is already a theory of perception and, inversely, that any theory of perception is already a theory of the body. Thus the complete overthrow of both versions of Cartesianism, empiricism, and intellectualism is implied. The outcome on the ontological level is the rejection of the psychophysical dualism which pervades the sciences of man and the affirmation of an existential ontology of embodiment as the root source from which any adequate theory of body and perception must arise. The lived body, the embodied subject immersed in a world pregnant with unwon significance, becomes the basic theme of the existential version of the primary perceptual situation. This unitary and relational ontology is held to transcend the dualism of Cartesianism.

The theory of expression in *Phenomenology of Perception* is situated within and patterned after and upon the theory of perception. It follows the same polemic pattern vis-à-vis the Cartesianism of dominant linguistic theory that the theory of perception follows in relation to psychophysical dualism. This is the Cartesianism which makes language psychophysical. "Physical" *sounds,* in the case of speech, are the pure body-matter of language, while "psychological" *meanings* are the mind of language. These two separate and distinct realms of linguistic being are related in some way in actual language.

While it is important to see how the Cartesianism of the linguistic sciences is isomorphic with that of other forms of psychophysical dualism, it is more important here to begin to discern the existentialist alternative concerning language posed by Merleau-Ponty. This alternative again parallels what was done with the theory of the body and perception. Language, expression, is ontologically always found *embodied:* "The word has a meaning."[4] Merleau-Ponty's whole theory of language is one of embodiment.

The argument is essentially the same as that used for the lived body: the concrete experience of speech and language is always first and primordially one of embodied meaning. A pure sound without significance is a construct; phonemes, morphemes, and the like, are conceptual "explanations" of sounds—they do not occur within concrete experience. But so is a pure thought! Both are the linguistic equivalents of sense data, the objective body, and the mental activity of the empiricist and intellectualist traditions. They are constructions which "explain" the ambiguities of experience. Linguistic "mechanics" on one side and pure "intelligence" which merely uses language as a tool on the other both fall short of noting the concrete incarnation of meaning which Merleau-Ponty offers as the existential alternative.

"The word has a meaning" is the linguistic equivalent to the embodied subject: "The meaning of words must finally be induced by the words themselves . . . immanent in speech" (179). Existent language is embodied expression, while both a presignificant sound and a postlinguistic thought are dualistic constructions. Embodied, expressive meaning thus parallels the embodied, perceiving subject.

This theory of embodied meaning is, I believe, the central focus around which the usually noted features of Merleau-Ponty's language theory radiates. Both (1) the "return to the speaking subject" and (2) the "primacy of speech" usually remarked upon by commentators are dependent in their significance upon the central notion of embodied meaning.

Furthermore, what must be seen as inextricably tied to the key of embodied meaning is the method of phenomenological genesis which Merleau-Ponty sees as his nuanced divergence from Husserl. The phenomenology of genesis is what creates the full *existentiality* of this phenomenology. Not only does it seek to "put essences back into existence" (vii), to think back to the lifeworld, but the weight given to genesis ultimately overcomes the sense of essence entirely. That "the greatest lesson which the reduction teaches us is the impossibility of a complete reduction" (xiv) is not so much a negative comment upon Husserl as it is the affirmation of what Merleau-Ponty understands an *existential* phenomenology to be.

This use of a phenomenologically described genesis *is* what both uncovers and justifies the sense of embodied meaning Merleau-Ponty elaborates. Symptomatically it is instructive to note the main class of examples which he uses to illustrate the phenomenon of expressive embodiment. The paradigms come from learning new meanings: (1) the child learning speech, (2) on first understanding others, (3) learning a new language, (4) learning a new philosophy, (5) the lover revealing his feelings, (6) the writer or philosopher struggling with awakening a sense of primordial experience, and finally (7) the mythical "first man who spoke" (178–79). These examples, adumbrated in varying ways, once again are illustrations of the stylistic movement of Merleau-Ponty's thought: the movement from an initial ambiguity through struggle toward a birth of new meaning. Thus for a third time we are plunged into the thematic problem of speaking in such a way that the new may be expressed. How does one give birth to the position which is phenomenology?

If a theory of embodied meaning is central—"the word has a meaning"—there remains a need to outline the configuration of that embodiment. Merleau-Ponty notes "there is thus, either in the man who listens or reads, or in the one who speaks or writes, a *thought in speech* the existence of which is unsuspected by intellectualism" (179). Thought-in-speech, to coin a language version of being-in-the-world, is the expressive dimension of human existence. What is embodied expression?

1. To begin, it is clearly a behavior or performance of the living subject. Language, in Merleau-Ponty's sense, is not something *had* by a subject, it is the subject in action. Speech in the broad sense used by Merleau-Ponty is the *performance* of thought: "Thus speech, in the speaker, does not translate ready-made thought, but accomplishes it" (178). The same performative emphasis is found in the interpretation of naming, that speech-act so often taken as central to linguistically oriented philosophies: "The denomination of objects does not follow upon recognition; it is itself recognition" (177) and "For the child a thing is not known until it is named, the name is the essence of the thing and resides in it on the same footing as its colour and its form. For pre-scientific thinking, naming an object is causing it to exist or changing it" (177–78). Naming *is* a performance.

2. The expressive activity of the subject in speech is intentional, directed, and focused activity. And as with all phenomenological intentionality, such an action is both internal and external, or better, already outside the enclosed self and directed toward the world. The performance of thought internally is as linguistic as actually spoken thought: "Thought is not 'internal' then, and does not exist independently of the world and of words . . . we can silently recall to ourselves, . . . through which we acquire the illusion of an inner life. But in reality this supposed silence is alive with words, this inner life is an inner language" (183). "Inner speech" is also embodied expression. Language is not private—nor is it public—it is between subjects, intersubjective, it is "a synchronizing of my own existence, a transformation of my being. We live in a world where speech is an *institution*" (183–84).

3. One can better say, rather than that man has or uses language, that man *is* language: "Language is much more like a sort of being than a means."[5] The thought-in-speech is a style of living in language: "The linguistic and intersubjective world no longer surprises us, we no

longer distinguish it from the world itself, and it is within a world already spoken and speaking that we think."[6] This thought-in-speech in which we live is itself quite concrete and even particular: "We may speak several languages, but one of them always remains the one in which we live. In order completely to assimilate a language, it would be necessary to make the world it expresses one's own" (187). Embodied expression is concrete and positional, the place from which one views the world. There is no metalanguage of disembodied meanings floating over and apart from actual languages: "If there is such a thing as universal thought, it is achieved by taking up the effort towards expression and communication in *one* single language, and accepting all its ambiguities, all the suggestions and overtones of meaning of which a linguistic tradition is made up, and which are the exact measure of its power of expression" (188). Thought-in-speech is embodied language. Thought *is* body in the same way that the subject is body.

If linguistic dualism is rejected, there is no realm of ideal and complete meanings above and apart from actual language, nor is there a realm of pure physical nonmeaningful motion which may be "used" by mentally employed meanings. And if all language in the human sense is existentially embodied, then the result is that the performances, positionings, and utterances we make are all made "inside" this existential language. Our being is being-*in*-the-world, here our thought is *in* language. But this poses an enigma: "One would have to know the language in order to learn it."[7]

Merleau-Ponty accepts this enigma: "This sort of circle, according to which language, in the presence of those who are learning it, precedes itself, teaches itself, and suggests its own deciphering, is perhaps the marvel which defines language" (39). Language does convey itself. Although there are several neat conceptual devices developed by Merleau-Ponty to justify this totality of engagement within language—for example, the first word of a child functions as sentence, the part is already a whole (40)—it is more important to grasp the internal movement which characterizes the birth of speech.

This movement, again the birth of meaning, is one which follows the pattern *from ambiguity toward clarity*. In Merleau-Ponty's phraseology this is the movement *from silence to speech*. Several important aspects of this movement need to be noted.

4. Strictly speaking there is no state of meaning prior to existential language. We have been led astray by those who have spoken of a "prelinguistic" state insofar as we have been led to believe that this state is equivalent to a state prior to meaning. There is, in Merleau-Ponty, the movement from silence to speech, but that is not a movement from

nonmeaning to meaning; it is rather a movement from the implicit to the explicit, from ambiguity already *pregnant* with significance to the expressed significance of speech. If meaning is "born" it is because the world is already pregnant with that possibility.

Meaning, in an existential theory of language, is the entire movement from silence to speech. It is a ratio of the implicit to the explicit. Thought, in the usual sense, is but a focus within the totality of meaning: "Because meaning is the total movement of speech, our thought crawls along in language. Yet for the same reason, our thought moves through language as a gesture goes beyond the individual points of its passage" (43).

5. It is within the context of this movement from implicit to explicit significance that the continuum of metaphors and behaviors which surround explicit speech takes shape. "Gesture," "gestural meaning," even the "indirect voices of silence" in painting and the concrete meanings of music which so well illustrate the incarnation of sound and meaning are all in between silence and speech. They are meaning-activities of the subject, but short of the explicitness of speech.

6. If the first movement is from implicit to explicit, from silence to speech, a reverse side of that movement remains implied in every speech-act. There is no final, no complete expression: "Now if we rid our minds of the idea that our language is the translation or cipher of an original text, we shall see that the idea of complete expression is nonsensical" (43). The existentiality of language is such that the field of implicit silence is always broader than the focus of explicit speech. In this sense speech is essentially finite even if also open and indeterminate.

7. It also means that the broad field of silent, implicit meaning lies as present background to all that is actually said. "Behind" what is said lies the unsaid. For Merleau-Ponty this is to imply that all speech, all explicit language, is *indirect:* "But what if language expresses as much by what is between words as by words themselves? By that which it does not 'say' as by what it 'says'?" "All language is indirect or allusive—that is, if you wish, silence" (45, 43).

The existentiality of Merleau-Ponty's language theory begins in anti-Cartesianism. In this it is part of the entire movement of phenomenology to counter the dualistic division of man and his world into matter and mind and instead to reassert the essential insertion of man within his world as incarnate being in a lifeworld. The existentiality of Merleau-Ponty's theory of language ends in the affirmation of embodied expression, thought-in-speech, and speech-in-silence, which attempts for language what is also attempted for the subject in the world.

All the various parts and aspects of Merleau-Ponty's theory of language may be seen as a Gestalt, a coherent whole, when viewed as the exemplification of a basic phenomenological model. For the sake of simplicity I shall use the terms *field* and *focus*. The field is the totality of presence which may be differentiated according to the question addressed to this totality of presence. Thus, if our question is visual, the field is the whole of the visual field before one. Focus is the region within the field which is attended to. Again, if the example is visual, the focal center may be a certain object which stands out against the background of the visual field. (I have characterized both here in noematic terms, but the noetic reflexiveness might also be characterized. For example, that which stands out must be correlated to my act of attending.) Extended, the notion of field is the phenomenological world; focus, the phenomenologically explicit attention within the world.

Silence, in Merleau-Ponty's use, is the field of pregnant, latent expressiveness always already present to the living subject. Pregnant silence is always and wholly present; man lives within the world of implicitly meaningful silence: "In short, we must consider speech before it is spoken, the background of silence which does not cease to surround it and without which it would say nothing" (46). *Speech* is the focal center, the explicit foreground of meaning which floats, varies, is directed but which always stands out against the background of silence. The clarity of explicit meaning is a relative clarity, relative to the implicitness of the background: "The clearness of language stands out from an obscure background, and if we carry our research far enough we shall eventually find that language is equally uncommunicative of anything other than itself, that its meaning is inseparable from it."[8] The movement of speech is a movement from and upon the field of silence.

This is also the reason why speech is the taking of a position. Focusing is taking a position within the field; it is a selection. But it, too, is relative. The position does not remove or obliterate the field, it merely allows it to remain background. The "indirection" of silence behind speech is not removable, and equally the explicitness of focus is not capable of completeness. If speech is necessarily focus, it can easily be seen that completeness of expression is impossible. Completeness belongs latently only to silence, to the unsaid.

The focus against the field may be exceedingly fine or broadened. Thus within speech itself—better, speech within language—the focus is the moving ray situated within the world of silent meaning: "Since the sign has meaning only insofar as it is profiled against other signs, its meaning is entirely involved in language. Speech always comes into play against a background of speech; it is always only a fold in the immense fabric of language."[9]

From the primordial field of silence to the most explicit attainment of thought-in-speech also lies a concentric and leveled set of stratifications. Gestural meaning, if "lower" than speech, is also broader. Within speech itself such stratifications also occur: "any linguistic operation presupposes the apprehension of a significance, but that significance in both cases is, as it were, specialized: there are different layers of significance, from the visual to the conceptual by way of the verbal concept."[10] But these stratifications are merely fine distinctions within the broader concept of field and focus which are called for to differentiate the degree and range of the focal center.

All expressivity, whether "higher" or "lower," more or less finely focused, is a positioning within the world of pregnant silence. Concretely, in man's case, thought-word-sound in existential speech is the region where expressivity is most clearly heard. The posturing which is speech is singing the world, and "singing the world" is now stronger than metaphor.

One may even speak here of expressivity as an "essence" of man in spite of Merleau-Ponty's demur concerning essences. Expressivity is at the least a dimension of being-in-the-world which takes distinctive human shape in speech. But it may be more than a dimension.

In Merleau-Ponty's works the question of language was imbedded in and followed from his discoveries within the perceptual world. Expressivity in *Phenomenology of Perception* took up but one chapter, even at that titled "*The Body* as Expression, and Speech" (emphasis mine). The other explicit essays, paragraphs, comments, also constitute but a small amount of the whole. Yet there is reason to believe that the reversed role of language in relation to perception taken here shows something at work in Merleau-Ponty's thought as well.

If expression is always embodied and if the world itself is that field of pregnant silence, then language in its broadened sense is not just one dimension of being. My thesis is that the result of taking language in the way Merleau-Ponty has is *to have made the question of perception enigmatic.* Furthermore, I believe that Merleau-Ponty in his later works had himself begun to reevaluate the role of perception. Perception, in *The Visible and the Invisible,* becomes the "perceptual faith." But faith recognized as faith is already doubt: "Philosophy is the perceptual faith questioning itself about itself. One can say of it, as of every faith, that it is a faith *because* it is the possibility of doubt."[11]

This doubt is such that perception and language are tied together more intimately, more inextricably than ever. If perception is "primary," it is not bare perception. Although Merleau-Ponty maintains his notion of inquiring back and down into the levels of experience—to the degree that he leaves himself open to those who see him seeking a level of prelinguistic meaning—that which is found is always sayable. Philosophy "asks of our experience of the world what the world is before it is a thing one speaks of and which is taken for granted, before it has been reduced to a set of manageable, disposable significations; it directs this question to our mute life, it addresses itself to that compound of the world and of ourselves that precedes reflection." But what it finds there is already open to speech: "But in addition, what it finds in thus returning to the sources, *it says*" (103; emphasis added).

What is at issue can be put quite simply if viewed tangentially in terms of another broadly accepted philosophical distinction. Phenomenology in its anti-Cartesianism claims to have rediscovered the living subject in a lifeworld. This rediscovery claims the power to overthrow the Cartesian notions of psychophysical dualism; mind and matter are to be transcended. However, a similar and no less tenacious distinction haunts our intellectual world as well, the distinction between *nature* and *culture*. Broader in its way, and thus perhaps more difficult to dispel, this dichotomy must fall, too, if phenomenology is to work out its program. Given this distinction, perception in its way belongs roughly to nature, while language belongs more clearly to culture. Thus language is "added" to nature.

There is evidence in Merleau-Ponty of an ambivalence with respect to the tradition of nature versus culture. In his essay "The Primacy of Perception," he appears to appeal overtly to the distinction;[12] in *Phenomenology of Perception* he speaks of speech as "the surplus of our existence over natural being" and links language to "a linguistic world and a cultural world,"[13] but increasingly the implication is one which must eventually call the nature-culture dualism into question. From the beginning he admits that human perception is different from animal perception, although without linking this to culture as such. But in *The Visible and the Invisible* perception becomes enigmatic precisely in relation to "cultural" factors. For example, in noting dramatic changes in the history of art regarding perspective, "I say that the Renaissance perspective is a cultural fact, that perception is polymorphic and that if it becomes Euclidian, this is because it allows itself to be oriented by the system."[14] Perception here is strongly relative to culture. But inversely, Merleau-Ponty notes, "What I maintain is that: there is an informing of perception by culture which enables us to say *that culture is perceived*" (212; emphasis added).

Nowhere, to my knowledge, does Merleau-Ponty make a sustained attack upon nature-culture dualism as upon the previous distinction of psychophysical dualism, but the ambiguity of perception, now tied more thoroughly to that of meaning, language, and culture calls for that attack to be made. But how? The answer is found in the guiding theme of the need for a radical language in phenomenology. If we are to "see" in a new way, we must be able to "say" in a new way. A language needs to be born.

Man as language has this capacity. His perceptual meanings (the *lifeworld* in the best Husserlian sense) are open to the creation of the new. Merleau-Ponty was on this track when he was untimely taken:

> If this paradox is not an impossibility, and if philosophy can speak, it is because language is not only the depository of fixed and acquired significations, because its cumulative power itself results from a power of anticipation or of prepossession, because one speaks not only of what one knows, so as to set out a display of it—but also of what one does not know, in order to know it—and because language in forming itself expresses, at least laterally, an ontogenesis of which it is a part. (102)

Phenomenology as philosophical revolution in its "linguistic tactics" is implicitly and purposefully radical in direction. It is precisely that which is taken for granted which must be uprooted and, if need be, overthrown. Cartesian clarity is sometimes at opposite poles with the *poesis* necessary for radicalness: "But from this it follows that the words most charged with philosophy are not necessarily those that contain what they say, but rather those that most energetically open upon Being, because they more closely convey the life of the whole and make our habitual evidences vibrate until they disjoin" (102). What Merleau-Ponty begins to seek explicitly lay earlier in latent form in both the phenomenological turn and in the question of embodied, expressive being. What must be said may be said, but not in terms of what merely has been said. What is called for, given the logic of Merleau-Ponty's thought, is "wild meaning" (155).[15]

6

Paul Ricoeur's Place in the Hermeneutic Tradition

Were one to identify the three primary twentieth-century philosophers associated with the most contemporary form of hermeneutic philosophy, there is little doubt that these would be Hans-Georg Gadamer, Martin Heidegger, and Paul Ricoeur. While each of these "giants" of hermeneutic philosophy occupies unique positions and takes unique perspectives upon hermeneutics, each also belongs to certain strands of twentieth-century forms of hermeneutics which bring them into proximity.

This essay will not be a comparison, per se, of these three figures, but, instead, a more concentrated analysis of Paul Ricoeur's place within hermeneutic philosophy. It will situate Ricoeur within both a longer and shorter history of hermeneutics. Thus, to begin, the *troika* of contemporary, world-class hermeneuts may be seen to have certain common interests and trajectories: (1) Most obviously, in the cases of Gadamer, Heidegger, and Ricoeur, the most immediate common background lies in the prominence *phenomenology* plays in their versions of hermeneutic philosophy. Contemporary European hermeneutics is *phenomenologically informed.* (2) The emphasis which a phenomenological hermeneutic takes is ontological, at least in the limited sense that its aim is to produce a *hermeneutic of human existence.* And (3) contemporary hermeneutics, precisely in its *ontological* dimension, requires not only phenomenology, but also a hermeneutics which is deeply *historical* in its awareness. With Gadamer and Heidegger—who remain more closely paired with each other in some contrast to Ricoeur—the question of *tradition,* and particularly the philosophical traditions of the West, form an essential dimension to the hermeneutic quest, while with Ricoeur that same history is always also *religious* or *biblical* history Yet, whatever differences obtain between the members of this set, it also remains that a hermeneutic phenomenology with its perspective upon understanding more human existence sets it off from other contenders concerning interpretation and criticism on the Euro-American philosophical scene.

The above is a highly condensed depiction of the shortest history of contemporary hermeneutic philosophy. But contemporary hermeneutics belongs to a much longer history as well, and it is Ricoeur who repeatedly and most thoroughly characterizes precisely this history.[1] What follows is dependent upon his explication of that history although phrased in my own terms.

For my purposes here the longer history of hermeneutics may be seen to have had three phases. (1) The oldest phase links hermeneutics, not so much to philosophy, as to *biblical exegesis and interpretation.* The Bible as the primary sacred text forms the focus of ancient and medieval hermeneutics. Modes of interpretation thus presume (a) a text, (b) need of interpretation and with it the presupposition that some nonobvious meanings must be recovered, revealed, exposited, and (c) an already chosen tradition which is in some way privileged—in this case the religious history which surrounds the central sacred text(s).

While I shall not detail here the specifics of that tradition, specifics which include such items as layered levels of interpretation (allegorical, literal, symbolic, and other meanings), the implicit needs to relate across different historical times and even cultures, I do want to highlight several interesting features of premodern hermeneutics.

First, as a form of exegesis, biblical hermeneutics remained in some sense *distant* from either philosophy or theology. While philosophies and theologies might well inform the principles of interpretation, that mode of informing remained *indirect.* Second, the immediate context of a biblical hermeneutics was not essentially ever an *externally* critical hermeneutics—it took for granted a certain history of salvation. This factor, often indirectly carried over into modern hermeneutics, often was critiqued as a weakness within the hermeneutic tradition concerning its capacity as a critical tool for philosophy. Does a hermeneutic approach *necessarily imply a privilege to some particular tradition, religious history, textual meaning?* And, third, while so much of Western or European philosophical and theological history could be interpreted as a blend, contrast, or even contest between often discordant ancient cultural roots—those arising on the one hand from Greek and Hellenic roots and on the other from ancient Judaic and Christian roots—biblical hermeneutics tended to emphasize the biblical or Judeo-Christian side of that double tradition. In one sense this is to say that the tendency of philosophy to rely more heavily upon metaphysical categories, often implicitly ahistorical or essential, gave way in hermeneutics to a much more historical, event-centered approach. All of these factors will be seen to play an important role with respect to Ricoeur's position within the *philosophical* adaptation of hermeneutics.

2. What I shall call *Modern* hermeneutics actually takes its shape in the nineteenth and even in the early twentieth centuries, thus in terms of the usual periodization of "Modern" philosophy (seventeenth and eighteenth centuries) it is somewhat late in arrival.

Modern hermeneutics, again frequently characterized in similar terms by Ricoeur's large number of essays on precisely this history, both *expands* hermeneutics from its previous, more narrow religious focus and *transforms* hermeneutics into a method which determines the shape of the social (or "human") sciences and the humanities. Ricoeur identifies Friedrich Schleiermacher and Wilhelm Dilthey as the two primary shapers of Modern hermeneutics.[2]

Before and behind "Modern" or nineteenth-century hermeneutics lay the early Modern development of the *sciences,* primarily the natural sciences. In terms of what hermeneutics had taken as a central focus, this development could be seen as (a) a shift in subject matter interest, from the focus upon human and divine history to one upon "Nature," (b) a shift in method, from textual exegesis to a kind of immediate observation in conjunction with inferential or rational processes, and (c) a shift from what were previously religious to secular or "humanistic" emphases, particularly within early modern philosophy.

Modern hermeneutics accommodated itself to all these shifts and in the process broadened the concept of what hermeneutics could cover. Placed within this context, Schleiermacher may be seen as a transitional figure. Although his interests continued to be religious, he saw the need to make of hermeneutics a more *general* and *philosophical* method. Hermeneutics was to become a set of general methods for any texts or historical materials, a general, philosophical method. Furthermore, as self-reflexive, hermeneutics became the specific mode of philosophical self-understanding. In the process, of course, the result was a more "humanistic" version of hermeneutics. Broadened thus, freed from the narrower focus upon sacred texts, hermeneutics nevertheless remained primarily text and history oriented but also begins to become understood as a method for understanding the "human" as such—it begins to be "existential."

Dilthey's subsequent development also is directed at the early modern formation and challenge from the successes of the sciences. His influential distinction between the social or human sciences—those employing the methods of *understanding*—and the natural sciences—those employing the methods of *explanation*—formed the field of much subsequent dispute and discussion. Hermeneutics, here clearly identified with the humanities and human sciences, contrasts with the methods and approaches to the phenomena of nature which belong to the now autonomous methods of the natural sciences.

3. Contemporary or twentieth-century hermeneutics, here identified with the three philosophers noted above, extends the modern development even further, through the adaptation of Husserlian phenomenology on the one hand, and, on the other, interpreting the task of hermeneutics as that of providing an *ontology of human existence.*

Again, in keeping with the brevity needed here, I shall not extensively review all the aspects of phenomenology which eventually became relevant to contemporary hermeneutics (Ricoeur himself provides more than a score of articles in precisely this adaptation), but one does need to note that while the Husserlian aim, "to the things themselves," referred to making objects as present within experience and in relation to that experience, the process of "questioning back" (*Rückfragen*) functionally provided a kind of archeology of meaning for a possible hermeneutic.

Contrary to the misinterpretation of phenomenology as taking as "given" the objects and events of ordinary experience, phenomenology was a means of shaking loose the sediments of ordinary or "natural" beliefs surrounding such objects and events and of showing, retrogressively, how taken-for-granted contexts often *hid* more complex and primordial significations. The whole process of the "reductions" in their many forms were techniques for "deconstructing" precisely such ordinary belief sediments. By presenting phenomenological methods in this light it is easy to see how what Ricoeur calls the "grafting" of phenomenology to hermeneutics is possible. If phenomenology is the archeology of getting back to the "things themselves," hermeneutics is the archeology of unlayering meaning-sediments originally associated with texts, but to become a broader unlayering of *philosophical traditions.*

Historically, in terms of our three figures, it was Martin Heidegger who first initiated this "graft" of phenomenology and hermeneutics, particularly the early Heidegger in the period of *Being and Time* and *Basic Problems of Phenomenology.* The "destruction of the history of ontology" was a *phenomenological hermeneutic* of the entire metaphysical tradition, a questioning back to a kind of "origin" in the birth of Greek philosophy. In this one has a kind of hermeneutic which looks like a gradual unpeeling of layers of traditions back to a source, and while that was part of the Heideggerian project, it also contained a more critical moment as well. The origin was not something simply to be returned to, rather, it was to be seen as an origin which simultaneously "reveals *and* conceals." The Greek moment was a particular moment, one mode of Being revealing itself *historically,* but precisely by being seen in this way it is seen also as a limitation and as having a possible "end." Thus the "end" of a Greco-European tradition and trajectory could be thought.

If phenomenology provided a kind of deconstructive archeology for this project, it also provided, for Heidegger, a more explicit focus upon what Ricoeur so often calls a direct ontology of the human (Dasein). The privileged entry into the quest for Being, in *Being and Time,* was through the analysis of Dasein. And it was a "structural" ontology of Dasein which formed the core of *Being and Time,* thus the "direct" ontology which Ricoeur was to critique. It should be noted that the structure of Dasein was itself *temporal* and revolves around the exstases of temporality in which a projected possible future gathers a particular perspective upon a past into a then finally constituted present. Presumably into this "structure" all of the variants upon histories of Being take their specific shapes.

Gadamer, in a somewhat ironic way, begins from both this Heideggerian version of hermeneutics and from his own version of ancient, particularly Platonic, thought. The irony is that Gadamer's approach turns out to be simultaneously more "conservative" than Heidegger's, but also potentially more "liberal" in the central notion of multiple and converging *horizons.*

One essential trajectory which emerges from *Truth and Method* and which is maintained throughout the Gadamerian corpus is the demonstration that a philosophical hermeneutic, to have any weight at all, *must be prejudiced.* That is, it must assume or take a structured perspective, take upon itself a *tradition.* In one respect this claim repeats the Heideggerian insight that all revealing is simultaneously concealing and that any event of Being must reveal itself in this showing/occulting. But in another sense, the Gadamerian version of this claim is that such a revealing/concealing is *horizonal* and thus recognizable as in a deep sense both unique and limited. It implicitly grants the possibility of *other horizons,* thus opening the way, more explicitly than Heidegger, of what today might be called cross-cultural or multicultural horizons.

Put negatively and in the worst light, the Heideggerian claim that *philosophy* could only be thought in Greek and German (and one must not trivialize this claim as such since it refers to the way one can be in and see from language which is a particular "house of Being") yields in Gadamer to the possibility that there are other tongues which could be horizonally equivalent. In this respect there is a distinctly *postmodern* anticipation within Gadamerian hermeneutics. And this makes one central issue and problem for *Truth and Method* the question of how such horizons may fuse, or interplay with one another. Gadamerian hermeneutics is more congenial to both a linguistic and cultural pluralism than is Heidegger's.

I have posed the depiction of contemporary hermeneutics in precisely this way to show now the unique and even more contrasting way in which Ricoeur attacks the same set of contemporary questions. There are several ways in which Ricoeur's uniqueness might be posed; I shall begin with what may be called a long-standing and particular *hermeneutic methodology* which emerges from his earliest and persists to his most recent works. This is a *hermeneutic dialectic.*

To the best of my knowledge, all of Ricoeur's interpreters have recognized the extraordinary generosity he shows to all the alternative positions which he interrogates, and these positions have been many. They have included the "critical hermeneuts of doubt," Freud, Marx, and Nietzsche; they have included several variants of the structuralisms; they have included the Anglo-American traditions in philosophy; and in more recent times they have also included a wide variety of interdisciplinary approaches including historians, biblical critics, literary theorists, and the critical theorists (particularly Habermas).

In all of this, Ricoeur is a most voracious and careful reader. His intellectual others are always respected, taken seriously, and interpreted in ways which most readers find to be more "objective" than not—but, in the end, all these others also are "taken into" the Ricoeurean corpus itself. This is so much the case that some critics of Ricoeur think him eclectic, while others take him as a kind of latter-day "Hegel." And, certainly in the later works (for example, the large, three-volume *Time and Narrative,* collections such as *From Text to Action,* and his closest foray into Anglo-American philosophical themes, *Oneself as Another*) the speed and rapidity of this ability to devour has increased.

Deeper than the generosity and seriousness of his reading spirit, however, is a much more persistent pattern of thought. His version of a hermeneutic itself is *dialectical* in that Ricoeur holds that any single-perspective approach will always remain insufficient to penetrate the insights needed, and particularly for such topics as self-understanding or human ontology.[3] The gradual development and maturation of this approach can be seen throughout the Ricoeur corpus.

Although the shifts in this dialectical approach are nuanced, they may be detected in the way one central variable, *phenomenology,* gets played within the dialectic. In Ricoeur's earliest, although already mature, period, the overarching project was often called a "philosophy of the will." In terms of the original French publications, this is the decade roughly spanning 1950–65. This is the period of *Freedom and Nature, Fallible Man,* and *The Symbolism of Evil.* Within this trilogy—and reflected in the then contemporary associated books of essays such as *History and Truth*—it was clear that phenomenology played a central and focal role vis-à-vis the already interrogated "others" which were various forms of objectivistic studies.

In *Freedom and Nature* Ricoeur proposed to take phenomenology down to the level of the "bloodstream" itself and in the process used objectivist studies as a *diagnostic* to awaken a sense of fringe phenomena which a straightforward and simply Husserlian-styled phenomenology might overlook. But, in this period, it remained clear that the phenomenological core of Ricoeur's favored approach remained privileged, and it was to phenomenology that the other methods reflexively related. This period was one which also carried a more explicit "Hegelian" moment as well. Phenomenology, plus some form of nonphenomenology, produced a "third term" which was an enriched phenomenology. And while this echoes a kind of "Hegelian" dialectic, it was not and has never been a *totalizing* move within Ricoeur's thought. For Ricoeur any "Hegelian" moment has always been restrained by a "Kantian" set of limits, and Hegel remains played off against Kant throughout the Ricoeur corpus.

In what I shall call his "middle period," again in terms of the dates of the French publications, roughly 1965–75, two nuanced shifts can be detected. This is the period of *Freud and Philosophy, The Conflict of Interpretations,* and *The Rule of Metaphor* along with other essay collections of the time. The first shift, not seemingly radical but portentious, was a strengthening of the "other" from the phenomenological pole. It may be seen in the strong emphasis given to "the hermeneutics of suspicion" (which Freud, Marx, and Nietzsche exemplify) to *chastise* and discipline any "hermeneutics of belief" (which is "phenomenological," but also that which retains Ricoeur's always present, if subtle, religious sensibility). One material concern throughout Ricoeur's corpus is what may be called a theory of the self or of self-understanding. And while even in the early period Ricoeur denied the possibility that there could ever be a *direct* mode of self-understanding or self-transparency—and thus *no direct existential ontology*—one can see in the middle period even deeper doubts being placed upon the privilege of phenomenology.

The dialectic is deepening in the sense that a diagnostic set of clues which could be recalled phenomenologically now give way to the deeper need to take that which is *contrary* to the direct, the self-transparent, into account. The various forms of "false consciousness" show what may be an index not just for deeper self-understanding, but for the very contrariety within consciousness. There is a sense here that any self-awareness is not simply limited, but simultaneously self-deceptive, and thus it must call for a deeper therapy from the *other.*

The second nuanced change in this middle period, again not lacking in the early period but heightened in the middle period, is a turn toward a greater emphasis upon language. As early as *The Symbolism of Evil* Ricoeur held that even the most primitive "confession" of evil already occurs in language, but by *The Rule of Metaphor* the emphasis is stronger, and the possibilities of imagination which project whatever is possibly human are now essentially seen within the metaphorical structures of language.

In the middle period, however, there is one set of dialectics which continues to privilege a uniquely *hermeneutic* position which should be noted and refers to the diverse set of interrogations of structuralism which occurs in this period. Ricoeur's primary critique of structuralism revolves around its functioned privilege of synchrony over diachrony, to structure over *event.* Interestingly, in this context it is frequent that Ricoeur appeals to biblical history to make his point: not to either theology or philosophy per se, but rather to the critical studies of biblical thought (even if the latter is often called "biblical theology" by some). Both structuralism and what would today be called "foundational" philosophy are displaced by an "event"-oriented biblical-styled history.

In more recent or "late" Ricoeur, we once again find a set of nuanced shifts regarding his deeply sedimented dialectical habits (here I refer to the 1975 to present writings which include *Time and Narrative,* the self as other, and such collections as *From Text to Action,* most of which appeared in the eighties in French). The structure of *Time and Narrative* is instructive for this period.

Again we find that there is a dialectic concerning time. "Phenomenological" time (characterized by the principals, Augustine and Husserl) is played off against "cosmological" time (Aristotle as forebear). These two perspectives on time are traced throughout Western philosophy, into the times of both history and fiction, and finally not only are not resolved as to the *aporias* which the disjunctions display, but are left to remain a nonsynthesizable mystery. And even though there are mentions of a "third time," it is recognized as nonsynthetic. At most what "unites" the two times is a certain narrative (again a language expressed) strategy within which the imagination views human possibilities. Narrative, a linguistic form of larger unit than metaphor, as a kind of story is, of course, an essentially *hermeneutic* device. And it echoes, as in the middle period, a reflection of precisely the event-oriented biblical version of history which Ricoeur favors.

The most recent set of nuanced shifts can again be detected by the role the variable I have chosen now plays: phenomenology. Phenomenological time (phenomenology) plays one contrasting focus to a second, cosmological time, but in one sense both phenomenological time and cosmological time are "direct" and "structural" at least insofar as they are describable aspects of time. Narrative, a linguistic strategy, is essentially *hermeneutic* even in contrast to its phenomenological relations. Narrative is an event-history which is simultaneously "metaphorical" and thus *indirect*. One may note in passing that this angle on recent Ricoeur also places him amid the very contemporary and postmodern sympathies for "conversations," different "readings," and "worlds of the text." At the same time, the seriousness with which Ricoeur takes both his interlocutors and other disciplines never trivializes this result.

Taken to this point, Ricoeur's "dialectical" hermeneutic reaches a somewhat different result than might have been projected from his earliest period. On one side the dialectical elements of a binarity in opposition are heightened. If one takes the essays as condensed versions of the larger strategies, then not only do collections such as *From Text to Action* reveal the deeply sedimented habits of binary oppositions—ethics versus politics, ideology versus utopia, discourse versus action—but one may see the approach as almost formulaic. Yet it does not so abstract the interplay as to oversimplify, and the binarity is never a set of equivalences. There is a complex matrix of differences which, when it fits, does so in terms of a mosaic-like result. This is clearly *not* "Hegelian." (At the end of *Time and Narrative,* Ricoeur claims much more definitively than previously that Hegel must be superseded—although throughout all of Ricoeur's works Hegel is always played off against and limited by Kant.)[4] And, even if the pattern is a repeated one, it often yields rich insights as well.

If now one returns to both the short and longer histories of hermeneutics, Ricoeur's place can be demarked. First, with respect to his intellectual peers, Gadamer and Heidegger, one notes as Ricoeur himself does that his version of hermeneutics is more *indirect* than theirs. And, if less conclusive in some respects, it is also more open than the others. This is particularly the case with respect to the other, nonhermeneutic combatants on the field. Ricoeur's interdisciplinarity is thus a strength, because one can find more conversation between Ricoeur and his others.

One way of putting it is to say that with Heidegger, for example, one must fully enter into Heidegger's world. This is not easy as most readers of Heidegger know, but, on the other side, once one sees the world as Heidegger does, there is a kind of richness to the "world" seen in precisely his way, and Heidegger is very seductive. Ricoeur's world, on the other hand, has enough commonalities with many others that it is easier to enter: it does not have a single or single-minded opening, but rather it has many openings and is thus more communitarian.

Gadamer's world falls between the differences described between Heidegger and Ricoeur, but it is also a very "Western" or Eurocentric world. Its respect for tradition is one which mutes the revolutionary. And even in the fusion and interplay of horizons, the Gadamerian world is potentially one which includes and which eases the transitions. Ricoeur's world does not reduce or ease such transitions. Its very nonsynthetic character respects differences, although the necessary result is also one in which the rough edges remain, in which resolution is more hoped for than achieved. It is one in which, even within oneself, multiple voices remain.

To this point I have forefronted the variations upon Ricoeur's distinctive *dialectical* version of hermeneutic philosophy. But in the background and often showing through is the theme of a unique emphasis of a biblically styled historicity which also marks Ricoeur's hermeneutics.

One may note this again with contrasts to his peers, Heidegger and Gadamer. In both there is little doubt that a particularly Christian history also lies behind and within their versions of hermeneutic philosophy—but with a difference. Heidegger looks backwards and does his deconstruction of *onto-theology,* which is to say, that the Judeo-Christian dimension of the Western tradition which he interrogates is already "translated" into philosophical—and thus essentially Greek-Hellenic—frameworks. Once again this tends to make Heidegger's approach more "insular" and particular to one, albeit dominant, strand of even the Western tradition.

Gadamer, too, portrays, although less markedly, the same tendency. However, "traditions" are somewhat more broadly conceived and, as already noted, potentially seen to be open to a greater plurality of horizons. Gadamer's greater preference for Plato (over Heidegger's for Aristotle) and his deep background in legal (and hence more Roman traditions) hermeneutics evidence this shift to a somewhat broader approach.

Ricoeur, however, draws very deeply from the Judaic and early Christian traditions, as well as the tendencies of early hermeneutics, often in distinct *contrast* to onto-theology or more philosophically embedded modes of thought. What I have called his biblically styled historicity enhances precisely that heritage which was so often a contrasting strand to philosophical rationalism in Western thought. And although transmitted through the more contemporary and critical approaches found in modern "higher criticism" approaches, the contrast to the already philosophically translated onto-theology is marked. Ricoeur's reference texts here tend to be scholars like Gerhardt von Rad, Rudolf Bultmann, and Jurgen Moltmann, rather than the voices of systematic or philosophical theologians. Similarly, one finds a strong emphasis relating to the historical disciplines with references to a range of historical criticism from Louis Mink to Hayden White. This is part of Ricoeur's both more explicit and open interdisciplinarity, but also that which gives his hermeneutics much of its multidimensionality.

In terms of the longer history of hermeneutics, Ricoeur's version is the one which most thoroughly reciprocates with the earliest traditions, those which centered biblical texts. The event-history which recurs within his strategies in the different forms that emphasis takes is nevertheless also a *critical* version of that hermeneutic. As in his middle period, Ricoeur remains not just a man of faith, but one of a "second" or critically chastised faith.

One way to measure Ricoeur's place in the hermeneutic traditions is to do so by placing his work, from one's own perspective as an "other." When this is done I find that his is a position of respect for one, but nevertheless also a position which contains a complexity which always yields surprises and from which one can learn.

7

This Is Not a "Text," or, Do We "Read" Images?

Magritte's enigmatic title "Ceci n'est pas un pipe," placed upon his "image" of a pipe, poses a problem not unlike the problem of the "Text" in the family of post-, a-, or non-Modern theories currently fashionable in Euro-American philosophical contexts. Clearly the "image" pipe is *not* a real pipe, yet the confusion which the title, juxtaposed with a realistic painting of a pipe, produces raises a wide set of questions worthy of surrealistic inquiry.

One might invert the same problematic as applied to this chapter: in some sense it *is* a *text,* and thus to claim it is not would be to perform or claim a kind of trickery which belies its mode of being-as-taken. But in Magritte's case it is the "image" which is not the (real) pipe. Thus one might begin somewhat analytically and note that there are a number of matters which need to be sorted out: (1) the presumed *referent,* an actual pipe—but which "refers," the title disclaiming reference, or the image with implicit reference? (2) if the latter, it is a *representation,* or *image* which somehow representationally refers; and (3) then there is the ironic disclaimer found in the sentence itself, "This is not a pipe."

But this will not do precisely because the very field in which *referents, representations,* and *various kinds of metaphorical structures* obtain is also that which is under contestation in post-, a-, and non-Modern contexts. This family of contesters of modernism has both a positive and a negative side. Its positive side might be called its "textualism" or "discoursism," and it finds itself a proponent of a kind of metaphorical totalization in which the phenomena of *reading, writing,* and *texts* are spread out over the entire social and cultural "worlds" to be analyzed. Everything becomes a "text" or "textlike" with the characteristics of indeterminacy, arbitrariness, and finitude. It could be called in a significant sense a "literarization" of the World. The "reverence texts" which themselves help establish this mode of analysis are primarily those stimulated by and arising out of Foucault, Derrida, and Lyotard in the fashionable Euro-American discourse, but adumbrated in both very widespread "social constructionism" approaches as well as a "bodies" approach particularly developed by late-twentieth-century feminism. The negative side is the critique which addresses what are sometimes identified as Modern stances regarding a number of related phenomena. Let us look briefly at what is being contested.

First, any "realistic" theory of reference is contested. From structuralism on, and particularly among the poststructuralists, reference or reality-referencing disappears in the notions that language refers only to itself or is a play of signifiers/signified within an arbitrary or, as I shall put it, literary fictive mode of construction. Positively, what emerges is some finite, indeterminate mode of "discourse," "episteme," or other largely socially or fictively constructed "textlike" phenomenon.

Second, similarly, "perception"—while frequently used in postmodern contexts—also is contested with respect to what I shall call "Body One," a located, sensory body. In its most radical form one finds the Derrida position as one which claims "there never was perception,"[1] in that the discovery of "presence" always takes place within some form of nonpresence which, in turn, relates back to inscription or trace or writing. In less radical form, "perception" is seen to be a Modern invention—along with Man—in Foucault's sense that perception is socially constructed and as such can be both invented and dis-invented in some present or future episteme. In short, the perception which remains is solely the cultural perception which is socially constructed.

Third, bodies, in my sense of "Body One" as a being-here, located, sensory being with specific styles of movement, are also contested, either in a conservative sense as being malleable down to a highly reduced biological dimension, or replaced as a social body whether in the form of the body of the condemned (as in Foucault),[2] or as a breasted being whose breasts are clearly socially constructed (as in Young).[3] In my language, "Body Two," a social body, substitutes entirely or almost entirely for one's "own" or "lived" body in the Merleau-Pontean sense.

Moreover, one must also recognize that the combatants have clearly gained territory, so much so that to try to return the contest to one in which "reference," "perceptions," and "bodies" simply regain previously lost ground would be seen as a retrogressive and thus defensive move. Thus I shall not take that tack. Quite contrarily, I want to take an example set which may be located precisely *between* the combatants, but one in which the issues of reference, perception, and body reemerge with a new and different kind of urgency for examination. That example set will revolve around what I shall call *image technologies*.

Image Technologies

One highly proliferative technological development in the late twentieth century has been the class of *image technologies* which span the entire social and cultural worlds from the most popular forms of imaging—for example, MTV—to the most exciting science research frontiers in which radical advances in medical imaging technologies are matched by equally new image technology advances in Earth and space sciences, in short, from the most micro- to the most macrophenomena of scientific interest.

This development has spawned, in my university, an interdisciplinary "Image Group" which contains radiologists, film theorists, and philosophers. We are learning some interesting things from one another —in part precisely because we bring to bear such very different interests and concerns. I shall begin with a few report-like remarks on some of our results.

Let us begin with an image technology far at the popular and nonscientific end of the spectrum covered: MTV. This style of imaging is largely a bricolage format in which some song is accompanied by a series of discontinuous, interfaded, highly suggestive image chains. Analysts are quick to point out that the layers of meaning are multiple, that there are subliminal "messages" being conveyed, and that the operations have a depth psychology, that is, Freud-Lacan et al., structure with both hidden and manifest, unconscious and conscious significations.

Note that here there is no question of any kind of "truth-functioning" or reality referencing, at least in any scientific sense. That is, the images do not transparently "refer" to some external event or reality as would be the case with a television monitor showing who is entering the door. Rather, if these images are in any way "textlike," they are like fictive texts. Fiction, after all, is not even constrained by any known set of physical, biological, or even social constraints. In fictive modes one can imagine speeds faster than that of light, cyborgs which do not exist, and stretch the limits of one's body beyond even a quasi-biological embodiment. This development is particularly extrapolated in virtual reality developments and displayed in such movies as *Lawnmower Man*.

This is not to say that the imaging itself is not constrained—it is constrained by the state of the art of the technology itself. Initially I want to take note that these constraints will pervade the entire spectrum of imaging technologies. For example, in what may seem too obvious and trivial, whether one is referring to MTV or to an MRI scan of the brain, the image-space is *framed space.* This implies that what is presented is presented as already distinct from ordinary or lived-bodily space precisely by the limited and selected-out *framing* of the image presentation. Additionally (apart from techno-fantasies which also appear in fictive form), this presentation is an on/off presentation. It is not the "involuntary" constancy of the on-so-long-as-conscious presence of ordinary experience. Thus, doubly, the framed and on/off (which is a temporal frame) nature of the imaging distinguishes it from the fulness of an embodied lifeworld. Here, already, there is an implicit metaphorical connection to *reading.* To watch MTV is to set aside time from ordinary actional life, as in picking up a book to read, and, moreover, to suspend actional "belief" and enter into, not the book, but the imaging progression. This carryover even overlaps the bodily insofar as the sitting-there is a partially disengaged mode of position.

Second, to date all mass image technologies also *lack depth.* (Holograms are, of course, exceptions.) Optical technologies yield, in television and cinema, a greatly foreshortened field for vision, and in such medical technologies as MRI, CT, sonographic, and other image scans, a mere *slice* or cross section of the imaged object (a brain, for example). And while contrivances are added to help overcome this limitation—such as the use of high-speed illusions using old-fashioned Renaissance perspective effects in science fiction cinema, or the additive technique of reconstructing more three-dimensional effects from multiple cross sections in medical imaging—the real-time limitations are clearly notable.

These effects are thus "reductive" when compared to plenary, constant, and active or full sensory experience. They are effects which keep imaging technologies as at most "virtual" rather than actually substitutable realities. But, were I to leave the analysis there I would thereby miss what is *additive* or *magnified* technologically in imaging. To locate this phenomenon I shall turn briefly to the more scientific use of imaging which, while using similar state-of-the-art technologies, does so within a very different set of trajectories and constraints from the popular end of the spectrum.

Needless to say, the "intent" of a scientific use of imaging does have a truth-function. It is essential that the image "truly represent" that which is imaged. Medically, the CT or MRI scan must show the abnormal growth to be there, and not that it be some artifact of the machine itself (this is always a technical problem in the development of and in the reading from new image technologies). But, interestingly, the way to attain that truth effect is not by means of anything like a "literal isomorphism" which a simple corresponding "representation" would give. (In older empiricist epistemology, representations could be considered to be correspondent if and only if they isomorphically replicated the thing being replicated. They were thought of as *copies* of the thing iself. Technologies *transform* all possible representations and are never purely correspondent. In this respect they are "hermeneutic.")

Within the multiple uses of scientific imaging, there is a spectrum which does run from partial isomorphism to variations upon isomorphism which vary away from the "literal" or copy form, toward a certain kind of "fictive" or technologically enhanced form of variation. A sonogram, for example, shows in real time the shape, movement, and configuration of a fetus (albeit in somewhat blurred and black-and-white display). One can recognize the body parts, see thumb sucking, and identify gender very early on. Moving to an MRI brain scan, one retains in a clearer fashion certain shape characteristics and thus spatial isomorphisms, but from a more sharply defined and now computer-enhanced set of contrasts. These can even be further enhanced by the use of "false color," which is also used in Earth satellite photography to highlight organic phenomena in contrast to inorganic ones. Note that even here, although the "intent" is to highlight in such a way as to reveal some "real" phenomenon, the complex techniques are very close to the fixing which can be done through digitally enhanced photography, which in journalism contexts today has become an ethical issue. What had been the implicit, although always naive, photographic claim for "realism" now gets deeply called into question. Hyperrealism in digitally manipulated photography is one variant away from isomorphism, but a strange one in that it is a variation which makes the thing look "more real" than it is.

Returning to scientific contexts, nonisomorphic variants today include such processes as light- or heat-enhancement techniques which clearly exceed normal bodily sensory capacities, but which nevertheless reveal "real" features of things. Militarily developed light-enhancement technologies allow seeing in the dark in ways which approximate feline capacities, and heat-enhancement photography can show the heat shadow of a fighter plane which just took off from a runway, thus revealing a close past in a dimension again approximating rattlesnake sensory capacity. Note that in this case, while none of the imaging described mimics old-fashioned copy-epistemology notions, it does, through variational means, "refer" to "real" effects.

Let us now move to a third use context for imaging technologies which lies between fictive and scientific constitutions: politics. In a series of documentaries undertaken by Bill Moyers, one program analyzes what is taken as "truth" in news programs. Moyers cleverly juxtaposes a smug Michael Deaver's claim concerning media manipulation with the news clips which were manipulated and with the reporter's responses to what had happened. In one such sequence the news clip carries the verbal news that President Reagan was about to veto a piece of legislation favorable to labor while the visual showed him happily drinking beer in a South Boston bar with workers. Deaver triumphantly proclaims that the pictures of "Reagan with the working stiffs" clearly overcomes and even contradicts the critical verbal report and that he knows that what is seen, not heard, is what will be "believed." The reporter, after the fact, concurs that, even when in contradiction, the positive, visual image is virtually always taken as "true" in contrast to the action which is merely verbally described. When this video was shown to our medical members of the Image Group there was what amounted to a gasp of recognition, but virtual disbelief that such overt manipulation of a news program could and did occur.

I will now complete my suggestive example-set with two more illustrations. First, the movie *The Capricorn Factor* humorously called into question one of the great technoscientific feats of the twentieth century—the first voyage of a human to the Moon. (Some may recall that some of the first television shots of the Moon landing were not, in fact, transmissions, but "simulations.") The movie—which actually echoed some cultures' beliefs rather than created them—shows in effect how the whole voyage could simply have been a media hoax. All of the shots of landing, the first steps on the Moon, the corny statement concerning "One small step for man, one giant step for Mankind," were portrayed as staged, thus undercutting again the implicit "truth" of imaging through television.

Second, and finally, there was the Gulf War, which in typical exaggerated fashion Jean Baudrillard claims "never existed."[4] Again on the news, a new style of technowar was displayed which featured "smart bombs" repeatedly going down ventilator shafts, antimissiles which were "shown" intercepting incoming Scuds, and other feats of precision warfare (repeated in the same clips over and over again to audiences which strangely both simultaneously believed and doubted the veracity of the presumed "realism"). Later analyses claim that instead of the claimed 95 percent Patriot accuracy of interception, at most 24 percent were successful and then often with disastrous results from falling debris from both missiles. And when, in the development of antimissiles, we learn that warheads were artificially heated to enhance the possibilities of a hit for demonstration purposes, the "cooking" of a technoscientific test causes doubt to border over from the political to the institutionally "scientific" itself.

My example set now has a spectrum, all of which utilizes image technologies, but set in contexts varying from sheerly fictive where no truth claim is made, to the political where truth claims are embedded in deep ambiguities of contestive and yet intermixed power plays and in which truth is very hard to recover, to the scientific uses which, under controlled constraints, nevertheless wish to make a strong truth claim, but which utilize modes of variation and enhancement which overlap exactly the previous two domains.

Applied to this example-set could be the similar spectrum of philosophical contests which are now occuring in the post-, a-, and non-Modern disputes. Overgeneralizing, the anti-Modernists tend to move the debate strongly to the left-hand side of the fictive-politics-science spectrum. All phenomena are taken to portray the socially and fictively constructed features of this domain. In part this echoes the cognate sensibilities and a shift of Euro-American fashionability themes away from the sciences (which had been the favored territory of the analytic philosophies) and toward the literary (with comp-lit now a virtual twin to continental philosophy). Thus when science is looked at as phenomenon, it is looked at as cultural institution, filled with both the imaginations of the fictive and the power plays of discourse-power, gender, race, and social construction which remains comfortable in the mid- to leftside of the spectrum. Much of the "strong program," feminist criticism, and other social constructionist analysis—and often with great insight—portrays science in this way. Yet, while a critic like Evelyn Fox Keller, characterizing deep gender differences both in science and in personality conflicts, (quoting Mary Ellman) quips: "Faced with the charge that 'women always get personal,' Ellman counters, 'I'd say men always get impersonal. If you hurt their feelings, they make Boyle's law out of it.' "[5] And, later extrapolating into noting that science results, in spite of gender and personalities, in generalizations like Boyle's law, "The fact that Boyle's law is not wrong must, however, not be forgotten" (11).

At this point it might be possible to see the image spectrum I have outlined as populated by association with three distinct social and theoretical communities: at its fictive extreme (MTV) there is no pretension toward truth or reference claims, and perceptions are simply imaginative variations without even sought-for structures. In the middle, the political or Foucault-like area, it is realized that whatever is portrayed, simply by virtue of the naive but powerful perceptual-bodily "belief" that "seeing is to believe," is therefore manipulable within some power domain or discourse or episteme, and thus social constitution is a kind of power-knowledge. This, then, leaves the far right of the spectrum to the scientific community which necessarily comes off as, if objective in any sense, clearly also somewhat naive with respect to what image transparency with respect to truth claims might be. Yet, interestingly, it is this community which is often the least naive with respect to the technical processes which actually yield the results, however isomorphic or enhanced, and the most sophisticated in guarding against and recognizing "instrumental artifacts." Therefore, instead of reverting to the easy result of taking technology—in this case the class of image technologies—as simply neutral and open to multiple uses depending upon the telos of the communities involved, I want to focus in more closely and more critically upon the human or community-technology relations.

Critical Instrumental Realism

In much of my earlier work on technology I differentiated between different types of human-technology relations, sometimes intimating that what I call *embodiment relations* are often preferred to others.[6] These relations are ones in which the technology not only becomes maximally transparent, but which quasi-symbiotically becomes a kind of extended embodiment (in my Body One sense here). Contact lenses, the well-crafted hand tool, but also larger and more complex technologies which allow us to get at the environment mediatedly, but through a perceptual-bodily ease, are all examples. In all these cases the technology is incorporated into perceptual-bodily actions as per the formalism:

(I–technology) ----▸ World

Image technologies, however, are interesting cases. In terms of my earlier distinctions, image technologies would ordinarily be thought of as *hermeneutic relating* technologies, although in particular cases retaining certain embodiment features. A hermeneutic relation is one which varies far enough from bodily isomophism (including both space and time factors) to be more "text" or "language"-like than body-like. Thus, for example, while a real-time television monitor directed at a bank teller's position can be thought of as a kind of technologically mediated and extended vision, its out-of-body position, its distance, and its obvious highly reduced field removes it in part from the immediacy of embodiment. Yet, like all image technologies, it retains and portrays—unless one is critical—its pretension toward "seeing is believing." We are all dramatically aware of this in the recent Rodney King and LA truck driver video cases. These uses retain a "body-like" seeing. But, in terms of my earlier work, most image technology fits better into *hermeneutic* rather than *embodiment* relational schemes. That is to say, the perceptual focus is *upon* the display screen, *through which* there is a presumed reference, as in the formalism:

I ----▸ (technology-world).

And for this very reason the question of "reading" images carries both positive and negative weight.

Image technologies, however construed, insofar as they retain perceptually isomorphic features also *presumptively* carry with them the aura of seeing-believability. The bank robber, viewed either in real time by the guard upstairs, or later in a more video-like repeat time display, or Rodney King shown on national news via the doubly transformed video/television display, carry this "eyewitness" quality. Here the continuum from the oldest photography to the latest complex manipulations of hyperreal imaging carry the same implicit seeing-believing claim. And this is the case *in spite of the fact that we know better.*

At bottom, this seeing-believability is tied to our basic and "primitive" bodily engagement with the World, our actional commitment which is echoed and mimicked through imaging technology displays. But, if this bodily-perceptual engagement is a constant, so also is its reflective, critical extension. The only cure for any perceptual illusion or error, as Merleau-Ponty so rightly pointed out, is more and and better perception.

That is why what is needed—indeed all across the spectrum noted —is a *critical instrumental realism.* That mode of realism is and should also be a *phenomenologically* critical realism which is provided precisely by *variational theory.* Let us then return to a few of the examples above and note what variations, within the contexts noted, might show.

Beginning with the science cases, there is an almost deliberate attempt to practice variations. My colleague Bob Crease recently reported on biomedical imaging technologies in *Science,* where he shows how, by deliberately varying and comparing very different imaging processes, richer results are obtained.[7] Different technologies and different processes are aimed at the same target (coherence interferometry utilizes faint reflections of light, optical coherence tomography uses infared radiation, MRI vibrates nuclei, etc.) and thus show nuanced differences which, almost like classical eidetic phenomenology, let the phenomenon show itself. Here, to get to a truth claim, the variational use seeks both precision and yet comprehensive variations.

Note now, too, how distant from the highly reduced "sound bite" culture of television is the above process if the analysis is left without variation. The Deaver manipulation of network news probably shows, underneath, the complicity of corporate politics with the media as much as anything, for what if the news had not "bought" the "working stiff" bit, but instead showed Reagan nodding off while the cabinet was in the process of determining what bill would be vetoed? (Critical variation paralleling the science example is, of course, possible in analysis as well, and the TV viewer could read more extensive news analysis from any number of sources as well, but here we return to the problem of informed, critical communities once again.)

Finally, in the case of MTV, it could be said that the artifacts of the technologies themselves become fascinating. The possibility of magically transmogrifying images, collage juxtapositions, nonsense streams, and the like become the very stuff of the "play"—is MTV deconstruction embodied? It is at least the playful use of imaging in fictive mode and in that retains the different kind of variation which belongs to creative and artful praxis.

I must, however, conclude: what I am suggesting through the examination of imaging technologies is that the roles of referentiality, perception, and bodies are not without import, but must be seen in contexts very different from their classical situation in earlier epistemologies. Referentiality results properly only from critical and "socially constructed" results within a trained community employing variational investigations. Perceivability is polymorphic and is always both bodily and cultural—no perception without embodiment, no embodiment without hermeneutic context. And bodies belong to more than one dimension, both "biological" here used metaphorically and "social" again in a metaphorical sense. The post-, a-, and non-Modern critiques thus call for a recontexting of all of these problems.

PART 3

ANALYTICS

In this section a sideways glance is made toward analytic philosophy, at least insofar as debates occur between philosophical subcultures. I show how even the choice and use of fictive examples contrast between analytic and continental philosophies. Hidden within these choices often lie deeper commitments concerning human bodies and technologies. Then, in a debate with Richard Rorty, who in many ways could be called an analytic hermeneut, I argue that a nonfoundational phenomenology actually solves many of the problems which show up in today's too highly textualized contexts. Finally, again turning back to technoscience, a set of questions is raised concerning how institutionalized science calls for a more highly visible type of critique.

8

Literary and Science Fictions

1. Teletransportation

> I enter the Teletransporter. I have been to Mars before, but only by the old method, a space-ship journey taking several weeks. This machine will send me at the speed of light. I merely have to press the green button. Like others, I am nervous. Will it work? I remind myself what I have been told to expect. When I press the button, I shall lose consciousness, and then wake up at what seems a moment later. In fact I shall have been unconscious for about an hour. The Scanner here on Earth will destroy my brain and body, while recording the exact states of all my cells. It will then transmit this information by radio. Traveling at the speed of light, the message will take three minutes to reach the Replicator on Mars. This will then create, out of new matter, a brain and body exactly like mine. It will be in this body that I shall wake up.[1]

Those who have carefully read Paul Ricoeur's *Oneself as Another* may recognize this as the more complete text of a sci-fi imaginative variation developed by Derek Parfit in *Reasons and Persons.* And those who read analytic philosophy will recognize this as a clever, but typical, fantasy using sci-fi literary examples rampant in this philosophical genre. Brain transplants, body teletransportations, mind-body switches—all play roles functioning not unlike phenomenological imaginative variations within this philosophical culture.

Parfit is identified by Ricoeur as one of his most formidable philosophical others in his attempt to meld into a "discipline [requiring] . . . a new alliance between the analytic tradition and the phenomenological and hermeneutic tradition."[2] Following his long and deep-seated habits of dialectically interrogating philosophical others—structuralism, psychoanalysis, critical theory, and most focusedly Anglo-American analytic philosophy—he produces what may be the capstone book of his career. For even though his dialectical and hermeneutic moves remain discernible in *Oneself as Another,* they are more subtly made, less visible in the foreground, and placed in the most complex and synthetic display of his philosophy to date. Ricoeur does continue to surprise me: he does not falter, he sharpens his tools, and he never fails to deliver new insights and challenges. And for those of us who have reached graybeard stage, we can note with envy and hope that Ricoeur has produced and published more books since he retired than in his career up to that time!

Is Ricoeur too friendly and easy on his critics and others, as Charles Reagan alleges in his new biography?[3] And is this particularly the case with his detour into analytic philosophy concerned with a theory of the self? Surely the sparse, virtually mono-dimensioned sense of analytic self-identity, identity in a desertscape, stands in sharp contrast to the richer, narrative, and polymorphic sense of the self which emerges in the rainforestscape of Ricoeur's theory. And one could, point by point, make these comparisons, but that is not what struck me in reading *Oneself as Another.* Rather, I shall turn to a crucial set of parallelisms which emerge in the fulcral fifth and sixth studies in the very middle of the book. For it is precisely in the Ricoeurean attempt to forge an alliance between analytic and phenomenological-hermeneutic traditions that these interesting parallelisms appear.

2. Two Philosophical Cultures

Ricoeur's take upon analytic philosophy—which includes discussion of Strawson, Ryle, and Davidson in addition to Parfit—is one which sees a symptom of two cultures in the differing uses of fictive, *imaginative variations.* Ricoeur begins by simultaneously identifying the different preferred literary, fictive examples as variations and by differentiating between "literary" fiction and "science fiction": "It will be one of the functions of the subsequent comparison between science fiction and literary fiction to place back on the drawing board the question of the presumed contingency of the most fundamental traits of the human condition."[4] So the first parallelism is the type and use of imaginative variations, in these cases exemplified in the two cultures by two different genres—literary and science fictions—each favored by one of the two traditions. This far I simply accept and agree with Ricoeur's insight into two parallel uses of fictive variations.

But, then, in a second move, Ricoeur surprises me and enters a territory about which he has seldom talked or worked upon. He identifies science fiction variations with *technology* (perhaps better capitalized as *Technology*):

> What [Parfit's] puzzling cases render radically contingent is this corporeal and terrestrial condition which the hermeneutics of existence, underlying the notion of acting and suffering, takes to be insurmountable. What performs this inversion of meaning by which the existential invariant becomes a variable in a new imaginary montage? This is done by *technology;* better, beyond available technology, this is the realm of conceivable technology—in short, the technological dream. (150, emphasis added)

It was, in fact, this identification which gelled my response, and now, departing from Ricoeur's stance, I will claim a second and much deeper background parallelism and argue that the two philosophical cultures are motivated—exemplified in Ricoeur and Parfait—by two contrasting *technomyths.* Ricoeur clearly recognizes Parfait's technomyth, but, I shall argue, he does not recognize his own. For deeper and broader than the preference for literary, fictive devices is the appeal to contrastive utopian and dystopian *technomyths.*

Deepest of all, however, is a third set of parallelisms, for embedded in the technomyths are two, radically different *body ontologies.* Here Ricoeur reemerges with clearer insights and by returning to the phenomenological sense of body distinguished from flesh shows how—in a new and more subtle way—analytic body theory might be called a "new" or neo-Cartesianism.

I shall examine each of these philosophical cultural differences in turn and then conclude with a suggestive deconstruction of the technomyths which constitute a hidden agenda between the traditions.

3. Analytic Sci-Fi versus Narrative Fiction

I return to Parfit's teletransportation: Parfit self-consciously worries, at least a little, that technologized fictions may extend too far. He notes that there is some hesitation and doubt within the Anglo-American traditions about the relevance and usefulness of such variations. Parfit points out that

> Some believe that we can learn little. This would have been Wittgenstein's view. And Quine writes, 'The method of science fiction has its uses in philosophy, but . . . I wonder whether the limits of the method are

> properly heeded. . . . [Such uses] suggest that words have some logical force beyond what our past needs have invested them with.[5]

Yet, for the most part, among those I shall call the "new Cartesians," sci-fi variations are the means of choice, now resonating with strong trends in popular culture with its cyber-cinema and "wired" embodiments.

If science fiction is the preferred mode of imaginative variations (to make logical points, to be sure), it is merely the foreground indicator of a much deeper background set of philosophical assumptions. I shall focus here upon two levels of these background assumptions, the implicit *technomyth* which seems to be functioning, along with a powerful *reductionist* notion of embodiment. Both of these background assumptions reinforce each other. But, first, a look at what Ricoeur calls the difference between "literary" and "technological" fictions:

> Literary fictions differ fundamentally from technological fictions in that they remain imaginative variations on an invariant, our corporeal condition experienced as the existential mediation between the self and the world. Characters in plays and novels are humans like us who think, speak, act and suffer as we do. Insofar as the body as one's own is a dimension of oneself, the imaginative variations *around* the corporeal condition are variations on the self and selfhood.[6]

In short, implied in what Ricoeur is calling the literary fiction is an embedded, implied theory of body as mine (me). Technological fictions—sci-fi analytic fictions—disrupt this ultimately phenomenological feature and attempt to make embodiment itself contingent:

> Are we capable of conceiving of (I do not say realizing) variations such that the corporeal and terrestrial condition itself becomes a mere variable, a contingent variable, if the teletransported individual does not transport with himself some residual traits of this condition, without which he could no longer be said to act or to suffer, even if it were only the question of knowing if and how he is going to survive? (151)

Clearly, the crucial difference, then, between what Ricoeur calls the literary fiction and the technological fiction is the implied sense of embodiment which the two variants display. Literary fictions, which Ricoeur prefers and illustrates in *Oneself as Another* in his discussion of *Antigone, Remembrance of Things Past,* and other works, naively accept and contain what I shall later discuss as the embodied self within a world and experientially centered variation. Technological fictions *displace* embodiment and do variations upon fictive nonexperienced (or nonexperienceable) possibilities. (As a minor caveat, I want to note that Ricoeur does to literary fictions precisely what Heidegger does to poetry: Were poetry only that which can "reveal Being," then much of what passes as poetry simply doesn't count, including most of what could be considered to be ironic, humorous, or satiric. Similarly, in Ricoeur the reduction is to the great novels, plays, and other literary vehicles which are primarily narrative in form, with distinct characters, voices, and identities. I wonder where Kafka's *Metamorphosis* or other nonhuman imaginations would fit which are clearly not distinctly sci-fi or "technological" or retain human corporeality.) So, here we have the first parallelism, two distinctly different styles of fictive imaginations, but variants which point to much deeper and broader cultural philosophical styles.

4. Contrasting Technomyths

I shall now move to the first background feature which distinguishes the two philosophical cultures. At this level this feature is one of an implied *technomyth,* or deep-seated and not always specified attitudes toward technologies. At the most extreme the technomyths take utopian or dystopian trajectories.

Analytic philosophers are rarely technological dystopians. In part, this should not be surprising given the historical associations of analytic philosophy with what was explicitly a central and highly valued appreciation of *science.* Not only was much of analytic philosophy to be an emulation of the "scientific," but it also brought with it a culture of the quasi-religious dimension of scientific utopianism. Science, if not now, then surely in the future, would make almost anything possible. In the contemporary situation, however, what was once "science" has become *technoscience,* science embodied through *technologies.* Repeating Ricoeur's critique of Parfit: "What [Parfit's sci-fi] . . . puzzling cases render radically contingent [the corporeal and terrestrial] . . . is done by technology; better, beyond available technology, this is the realm of conceivable technology—in short, the technological dream."[7] What Ricoeur is calling "the technological dream" is part of what I am calling a *technomyth.* In its analytic version it is utopian and fictive and holds implicitly that the conceivable is ultimately technologically feasible. And, in the contemporary situation, the technological dream often turns *biotechnological.* Thus the imaginings of transplants, teleportation, and mind-body changes have a technobiological shape and "causation." Nor can I too often point out not only that the slide from conceivability to future actuality barely conceals the utopian beliefs which go back all the way to Roger Bacon, Bruno, da Vinci, to reach a peak in the Enlightenment, but that these also are rampant in popular culture today.

The problem of critique here, however, does not lie with the task of distinguishing between the quasi religiosity of utopianism with what can become actual, but with a much deeper "contradiction" within which the "technological" remains concealed in science-fiction forms. This contradiction is deeply embedded in what I have heretofore termed *embodiment relations* in which humans relate to and through technologies as *quasi-bodily*.

It takes particularly poignant shape in the cases of prosthetic devices and at the edges of the technologies which fictionalized are termed "bionic." But the contradiction lies in a more general desire relating to all embodiment possibilities:

> There is . . . a deeper desire which can arise from the experience of embodiment relations. It is the doubled desire that, on one side, is a wish for *total transparency*, total embodiment, for the technology to truly "become me." Were this possible, it would be equivalent to there being no technology, for total transparency would *be* my body and senses; I desire the face-to-face that I would experience without the technology. But that is only one side of the desire. The other side is the desire to have the power, the transformation that the technology makes available. Only by using the technology is my bodily power enhanced and magnified by speed, through distance, or by any of these other ways in which technologies change my capacities. These capacities are always *different* from my naked capacities. The desire is, at best, contradictory. I want the transformation that the technology allows, but I want it in such a way that I am basically unaware of its presence. I want it in such a way that it becomes me. Such a desire secretly *rejects* what technologies are and overlooks the transformational effects which are necessarily tied to human-technology relations. This illusory desire belongs equally to pro- and anti-technology interpretations of technology.[8]

This applies, I believe, to Parfit's teleportation fantasies, and Ricoeur is right in calling it a "technological dream"—it is a fantasy exercised by the contradictory desire which both wants and does not want technology and which thus *hides* the technological.

If Parfit's agenda embeds its fantasies in technological utopianism, does Ricoeur's agenda do the same with respect to a technological dystopianism which for so many Euro-American philosophers also entails continued Romanticism? At no point in *Oneself as Another* does Ricoeur explicitly take a distinctly antitechnological stance, nor does he even clearly link his fictive devices with "technology." Indeed, the technological is almost invisible for his side of the cultural gap.

A version of continental Romanticism, however, does appear. It lies in what I shall call a "slide" from the phenomenologically established ontological relation between an embodied self and the environing world. Regarding literary fiction and its sense of embodied self, "In virtue of the mediating function of the body as one's own in the structure of being in the world, the feature of selfhood belonging to corporality is extended to that of the world as it is inhabited corporeally."[9]

Then, in a distinctly Heidegger-like move, Ricoeur identifies the phenomenological world with "Earth," which then gets toned down to the notion of *terrestrial condition*: "This feature defines the terrestrial condition as such and gives to the Earth the existential signification attributed to it in various ways by Nietzsche, Husserl, and Heidegger. The Earth here is something different, and something more, than a planet: it is the mythical name of our corporeal anchoring in the world" (150).

I shall not here follow, but simply assert, that in Heidegger the notion of Earth becomes a key variable in his critique of Technology, the calculative way of seeing the world which contains and belongs to Modern Science which is a tool of Technology-as-Metaphysics. In the "world-as-view," Earth can be degraded into a mere planet, but as a result of *technological* seeing. Does Ricoeur fall into this dystopian stance? I cannot make a strong case that he does, although his frequent use of "terrestrial condition" identified with a mythical Earth seems to suggest an affinity. The one clue that an implicit antitechnological trajectory may be part of a technomyth is his worry about a Parfit-like projection in which the sci-fi fantasy could become actual. Ricoeur interestingly argues that (phenomenological or) literary corporality carries something like a moral implication:

> An imaginary system which respects the corporeal and terrestrial condition as an invariant has more in common with the moral principle of imputation, [thus] would not any attempt to censure that other imaginary, the one which renders this very invariant contingent, be in its turn immoral from another point of view for the reason that it would prohibit dreaming? (151)

Caught now in not wanting to censure dreaming—technological fantasies —but at the same time wanting a morality which protects corporality and the terrestrial condition, Ricoeur cautiously worries: "It will perhaps one day be necessary to forbid actually doing what today science fiction is limited to dreaming about. . . . Let us simply express the wish that the manipulative surgeons in these dreams never have the means—or more especially, the right—to perform what is perfectly permissible to imagine" (151).

Ricoeur is not about to wish himself teletransported to Mars? But this caution falls short of the call to remain en-Earthed or to stay in the forest cottage of his German counterpart. Ricoeur's caution, however, insofar as it entertains a fear of technological possibility without recognizing the contradictory nature of the dream, remains in precisely the same dilemma as Parfit regarding the technological. I argue that technological utopianism and technological dystopianism equally hide the technological:

> The [contradictory] desire is the source of both utopian and dystopian dreams. The actual, or material, technology always carries with it only a partial or quasi- transparency, which is the price for the extension of magnification that technologies give. In extending bodily capacities, the technology also transforms them. In that sense, all technologies in use are non-neutral. They change the basic situation, however subtly, however minimally; but this is the other side of the desire. The desire is simultaneously a desire for a change in situation—to inhabit the earth, or even to go beyond the earth—while sometimes inconsistently wishing that this movement could be without the mediation of the technology.[10]

5. Different Bodies

We now arrive at the most crucial of the parallelisms lying in the *body ontologies* which are radically different between analysis and phenomenology. Ricoeur, in the tenth, ontological study, returns to the *phenomenological* sense of embodiment and the distinctions of Husserl and Merleau-Ponty between "body" and "flesh":

> To say that the flesh is absolutely here, and so heterogeneous with respect to any set of geometric coordinates, is equivalent to saying that it is nowhere in terms of objective spatiality. . . . And the "over there" where I could be if I transported myself there . . . has the same status of heterogeneity as the here of which it is the correlate. . . . [This is within an] *environing world,* taken as the correlate of the body-flesh. . . . The

> world as practicable completes fortuitously what has just been said about the internal, as it were, spatiality of the flesh.[11]

He also returns to the pre-extra-linguistic meanings of the flesh-world correlation:

> It is upon this prelinguistic relation between my flesh localized by the self and a world accessible or inaccessible to the "I can" that a semantics of action is finally to be constructed which will not lose its way in the endless exchange of language games. (325)

Flesh, phenomenological body, is *mineness,* experienceable. So, when Ricoeur turns to Parfit's body ontology, it must appear *reductionist.* And, while I would claim that the neo-Cartesianism of Parfit also reintroduces a "god's eye view" into his technological fictions, Ricoeur's critique is one which emphasized what is new in this revived Cartesianism:

> What the reductionist thesis reduces is not only, nor even primarily, the mineness of experience but more fundamentally, that of my own body. Thereafter, the true difference between the nonreductionist thesis and the reductionist thesis in no way coincides with the so-called dualism between spiritual substance and corporeal substance, but between my own possession and *impersonal description.* . . . The most radical confrontation must place face-to-face two perspectives upon the body—the body as mine, and the body as one body among others. (132, emphasis added)

Then, in one of the most thoroughly phenomenological descriptive arguments in the book, Ricoeur goes on to show how "This neutralization, in all the thought experiments . . . will facilitate focusing on the brain the entire discourse of the body." Here is part of his phenomenological description:

> The *brain* . . . differs from many other parts of the body, and from the body as a whole in terms of an integral experience, inasmuch as it is stripped of any phenomenological status and thus of the trait of belonging to me, of being my possession. I have the experience of my relation to my members as organs of movement (my hands), of perception (my eyes), of emotion (the heart), of expression (my voice). I have no such experience of my brain. In truth, the expression, "my brain," has no meaning, at least not directly: absolutely speaking, there is a brain in my skull, but I do not feel it. It is only through the global detour by way of my body . . . that I can

say: "my brain." . . . Its proximity in my head gives it the strange character of *nonexperienced interiority.* (132, emphasis added)

The body-as-brain, the reduced impersonal and nonexperienced body in Parfit, is adjoined to another favorite notion, the "cerebral trace" as the substitute for memory, another case of reducing the personal to the impersonal which *slides* the discussion over into the language games of "causal dependence" and other forms of physicalism still favored by analytic philosophers. Ricoeur concludes that "Parfit's fictions . . . concern entities of a manipulable nature from which the question of selfhood has been eliminated as a matter of principle" (133).

Perhaps, at this most extreme juncture, I should not call the divergent body ontologies a "parallelism" because there is much greater contrast here than in any of the previous comparisons. A body ontology in which "mineness . . . constitutes the core of the nonreductionist thesis" (135) and constrained by the phenomenological notion of body as flesh simply contrasts with the reductionist thesis which locates identity in brains, cerebral traces, and other impersonal, inexperienceable entities. Thus the parallelisms, instead of gradually converging, *diverge.*

6. Ricoeur's Analytic Detour

A detour presumably takes one on a side road but eventually must return to the highway—does the detour via analytic philosophy do this? Probably not. If not, why not? I think that the reason ultimately lies in the two radically different body ontologies which motivate analytic philosophy on one side, and hermeneutic phenomenology on the other. *Oneself as Another,* perhaps unintentionally, reveals this fissure. As I draw to a close, I want to observe that the two styles employ—in the services of their ontologies—two very different *constraint systems.* I shall contend, with some irony concerning the self-interpretations by these traditions, that analytic philosophy (here Parfit) operates with a *loose* constraint system, whereas hermeneutic phenomenology (here Ricoeur) operates with a *tight* constraint system.

At the level of fictive variations, it is easy to point out the looseness in Parfitian teletransportation fantasies. He begins with an easily imaginable transportation example: the "old method," a spaceship, is not only logically, conceptually, but also technoscientifically possible—"by today's lights," to employ a Quinean phrase. Such imaginings are implicitly constrained by the context of the best-known technoscientific parameters of the present. To be sure, he has reduced the time of a current probe from eighteen months to several weeks, a factor of *six*. The leap to the new method, the *Replicator,* radically loosens the constraints to the merely logically, conceptually possible with only minimal caveats to retain a "scientific aura." The Replicator still seems constrained to the speed-of-light speed limit of known science and seems to imply a biotechnology related to the beginnings of genetic, atomic, and molecular manipulations. The looseness of this constraint may be seen when the giganumbers implied in the fantasized technologies are noted. Currently it is possible to send a million bits per minute; for example, a single frame of an image from Mars contains ten-million bits and takes ten minutes to transmit. Were it possible to reduce the human body, with its ten thousand trillion cells, multiplied by the ten thousand protein units in each cell, yielding almost a billion trillion bits, and send these in today's technologies, it would take *750 days plus to get a body's information to the Replicator.* Clearly, this is no advance since it is slower than the space probe. But if we increase it by a factor of six, it still will take at least 150 days and still be slower than the probe.[12]

The very absurdity "by today's lights" of this situation only serves to sharpen further the notion that the thought experiment is a *wild* and barely constrained fantasy. Admittedly, Parfit does not take the even easier route of "Star Trek," which loosens the constraints far beyond the physics, let alone the technologies, of today. Parfit's Replicator does not have the capacity to go from "impulse" to "warp" speed and thus break the speed limit of the known universe. But the constraint system is largely open to mere logical and conceptual possibility. This leaves it far looser than any scientific constraint system which must take account of the known laws and patterns of physics, chemistry, biology, and *technology.*

Ricoeur's constraint system is much *tighter,* although it cannot be said to be a technoscientific set of constraints. It is rather a set of *existential* constraints governed by experienceable bodies in correlation with an experienceable surrounding world. His favored literary fictions also presuppose embodiment, temporality, and bodily spatiality. These are constraints which cannot "violate a constraint of another order, concerning human rootedness on this earth"[13] and which must contain the full sense of "mineness, which . . . constitutes the core of the nonreductionist thesis."[14] But if Parfit's constraints are too loose, then perhaps Ricoeur's are too tight. Humans have, after all, actually escaped—temporarily and in a technological cocoon to be sure—the Earth; and the perspectives which show Earth as (also) planet are not totally alienating since they show the finitude of our dwelling place and also begin to make it appear as a kind of integral whole. I even think that the ontological correlation of embodied self within a world might be able to include "worlds" if they are complex enough to have the ecosystems which are necessary to sustain us (we are, after all, not a solitary being, but a being-in-a-system). So, in my own way, I accept Parfit's rendering of myself and my world as contingent, although by a different route. But I suggest the dreaming which makes this possible needs to be constrained by a deeper recognition of the technological, taken existentially.

7. Teletransportation Redux

> I press the green button and fall unconscious. The worrisome dreams I had before this moment, about all the glitches early replicators had, when like computers they used to "crash" in the middle of a transportation, do not occur. I am, of course, unaware that the residual effects of a solar flare have passed by during my own weeks of transportation. I wake up and announce: [Read in falsetto] "I'm here; everything seems to be fine; I'm me. But there does seem to be something a little different . . ."

9

Response to Rorty, or, Is Phenomenology Edifying?

Response to Rorty

Were philosophers to have Academy Awards, and were reviews in the major journals and such national papers as the *New York Times* or the *New York Review of Books* nominations, then Richard Rorty's *Philosophy and the Mirror of Nature* would certainly have won the 1981 Oscar. For that book, with only one contender,[1] surely received more attention than any other philosophy book that year. The interest spread to Stony Brook, too, for in the fall term Patrick Heelan organized an informal seminar of some twenty faculty and doctoral students to read *Mirror.* I was, because of a schedule conflict, unable to attend but soon began to hear discussion in the hallways.

Bits and pieces began to take shape: "Rorty's decreed the end of analytic philosophy." "Rorty's turning continental." "Rorty thinks that the three greatest philosophers of the twentieth century are the late Heidegger, the late Wittgenstein, and Dewey." And, most often from my ACE[2] colleagues, "Well, its about time somebody from the establishment discovered what we've known all along about the state of contemporary philosophy." Still, I had not read the book.

Then the reviews appeared, repeating in many ways the above conversation in more academic prose. Some believed that with Rorty's strong emergent interest in *hermeneutics* evidenced in *Mirror* and with the repeated references to Heidegger, Habermas, Apel, and Derrida, that he had, after all, opened the way to continental philosophy. Moreover, his scathing critique of what some identified as analytic philosophy was unmistakable.

But it was not until early 1983 that my own schedule allowed me to read the book. I admit to some grudging reluctance at first, since, from the secondary information, I, too, believed that Rorty was a Johnny-come-lately to a perspective on contemporary philosophy which many of us arrived at fifteen or even twenty years ago! His choices of representative giants had even been anticipated in print far earlier, and again by William Barrett in his 1978 *Illusion of Technique.* Barrett chose two of the same individuals, although he substituted William James for John Dewey, but with much the same impact. Moreover, the death of Modern Philosophy, that is, the "foundationalism" of the Cartesian sort, had been decried by virtually every "classical" phenomenologist in all three varieties—transcendental, existential, and hermeneutic. Thus when I began to read, I received something of a surprise. First, the bulk of the book did not really have so much to do with anything like either a conversion to continental philosophy or a deathknell for analytic philosophy that the secondary interpretations seemed to emphasize. Rather, there was a reworking of a total perspective upon contemporary philosophy which in its most penetrating sense did not even distinguish clearly between analytic and continental forms.

Surely, Rorty's audience remained the AE, and his rhetorical and conceptual style remained clearly within those boundaries. I have already remarked upon the obvious invisibility of those who in ACE were already quite aware of the negative side of *Mirror*'s result. *Mirror* was, in the thrust of its attack, a kind of Kuhnian shift, a change of model or of categories by which one could interpret contemporary philosophy. It was a shift which, negatively, did claim that the end of Modern philosophy, insofar as it was a systematic, epistemological-metaphysical enterprise which constructed itself upon a *foundational* base, was no longer tenable. This was not exactly news to many of us—but what Rorty did further was to develop the thesis such that his emergent perspective which differentiated between foundational and edifying *hermeneutic* philosophy cut across *both* analytic and Continental fronts. Its result was, on one level, to undercut much of what had been taken for granted as differences between the two styles of philosophy and replace it with another.

I also found that I had to take Rorty at his word concerning what he was doing, for this attack upon foundationalism in both analytic and continental groups arose primarily from within analytic philosophy itself, from its more pragmatist sources. Rorty claims:

> I began to read the work of Wilfred Sellers. Sellers' attack on the Myth of the Given seemed to me to render doubtful the assumptions behind most of modern philosophy. Still late, I began to take Quine's skeptical approach to the language-fact distinction seriously and to try to combine Quine's point of view with Sellers'. Since then I have been trying to isolate

> more of the assumptions behind the problematic of modern philosophy, in the hope of generalizing and extending Sellers' and Quine's criticisms of traditional empiricism.[3]

The hard core of *Mirror* is exactly that. From within the larger analytic movement, Rorty has taken a pragmatist, antifoundationalist stance and argued that all forms of analytic foundationalism are untenable. This is clearly a severe attack, for it implies that the "science model" held by the early Positivists and retained through most foundationalistic philosophy must go.[4] And, if correct in a Kuhnian sense, this would also mean that much of what is taken as "normal" problems for analytic philosophy would not so much be solved or reworked, but simply disappear, become "uninteresting." This would be the case for much of the so-called body-mind problem as well. Philosophies are, of course, rarely responsive to refutations. Historically they tend either to be abandoned rather than die of rebuttal or, more likely, to undergo resuscitation by revision. Thus analytic foundationalism has in very recent years, reemerged as the New Realism Rorty refers to in his later *Consequences of Pragmatism.*

The internal attack, addressed to analytic philosophers, clearly arose primarily from Rorty's own readings of that tradition and its problems. This is clear in both the form and substance of the attack. Nor does it repeat the same kind of criticism made much earlier by both Husserl and Heidegger. Indeed, I suspect Rorty would think of their attacks as still within the foundationalist framework since it is possible to interpret Husserl as rejecting Cartesianism on behalf of transcendental idealism, and Heidegger's destruction of the history of ontology which covers over a more ancient and favored ontology as yet another foundation.

However, in one crucial way Rorty's attack does function like the earlier attacks of Husserl and Heidegger. Rorty's "paradigm shift," which resituates a perspective upon contemporary philosophy, is in practice something like the deliberate tactic of a "paradigm shift" employed by Husserl well before Kuhn. Husserl made such a shift of perspective an essential and deliberate part of phenomenology itself. This tactic, buried for some beneath his intricate machinery, is nevertheless exactly a purposeful shift of perspective. I refer, of course, to what Husserl called the shift from the "natural attitude" to the "phenomenological attitude," a shift which was both deliberate and fundamental for a different kind of "seeing" to occur. The elaborate steps of the *reductions* which appear in most of Husserl's works are the parts which go together or, better, the set of hermeneutic instructions which tell how to perform the shift. Unfortunately, too many interpreters simply either got lost in the intricacies or, worse, read Husserl literally.[5] What was important was to be able to experience what is seen differently, which, once attained, make the instructions either intuitive or unnecessary. (In a sense, Kuhn describes what happens in a shift, but how it happens remains for him largely unconscious. Husserl attempts to make shifting a deliberate procedure, a phenomenological rationality.)

The common perceptual model between Husserl and Kuhn is the *Gestalt shift*.[6] Gestalt shifts, those which take either ambiguous figures or grounds, are mini-illustrations of changes of perspective. Paradigm shifts or the shift from the "natural" to "phenomenological" attitudes are, of course, both more sweeping and more fundamental. But it is clear that Husserl made this phenomenon one central element of his own phenomenology. Rorty, with reference to the field of contemporary philosophies, practices such a shift.

Rorty's shift at its highest level of abstraction divided contemporary philosophies into those which are *foundational* and those which are nonfoundational. Then, within those which are non-foundational, he discerns a certain pattern which he terms alternatively *edifying* or *hermeneutic.* This is to say that the negative side of *Mirror,* the attack upon foundationalisms, is matched by a positive side, the development of a generic hermeneutic or edifying philosophy. It is here that the representative giants take their shape and role: late Heidegger, late Wittgenstein, and Dewey.

If each rejects systematic, hierarchical, structural, and foundational philosophical erections, particularly those of the Modern or Cartesian and Kantian sorts, then what is to be accepted also has need of clarification. For Rorty, language remains the guiding thread. In his preface, he indicates that what unites all he learned from the variety of his teachers and philosophical education was "I treated them all as saying the same thing that a 'philosophical problem' was a product of the unconscious adoption of assumptions built into the vocabulary on which the problem was stated—assumptions which were to be questioned before the problem itself was taken seriously."[7] This remains a central philosophical axiom with Rorty, and he has even produced a lexicon translating the main terms of *Being and Time* into vocabulary language.[8]

Language is a twentieth-century obsession for philosophers, both Anglo-American and Euro-American. But what is common to the giants Rorty cites is what may be called a certain *horizontalization* of language. Negatively, this is a rejection of hierarchies to language and specifically a rejection of the language/metalanguage developments of foundationalists. Wittgenstein, in the later "language game" and *Philosophical Investigations* period, of the three giants, was the most explicit about this connection. But the movement in Heidegger from the apparent structural foundationalism of *Being and Time* to the increasing hermeneutic horizontalization in his late work follows the same trajectory.[9] And, Dewey, although not directly concerned with language in the same way, unites both in his deconstruction of the sciences into human problem-solving activities which, again, are horizontalized.

Horizontalization implies negatively that there are no privileged language games, no disciplines, no privileged activities. There are only appropriate or inappropriate contexts and a diversity of fields. What seems as a kind of democratic anarchism here Rorty seems to apply to the very profession of philosophy. It must take its place as one type of conversation among others. Certainly, philosophers can no longer pretend to be either the overarching thinkers of the past or cultural mandarins with a higher authority. They can be *edifying* in the sense of a moral concern:

> The only point on which I would insist is that philosopher's moral concern should be with continuing the conversation of the West, rather than with insisting upon a place for the traditional problems of modern philosophy within that conversation.[10]

This is a modest view of philosophy, an essentially communicative, interpretative one. But Rorty intends it to be positive within this modest position; his choice of edifying is deliberate: "Since 'education' sounds a bit too flat, and *Bildung* a bit too foreign, I shall use 'edification' to stand for this project of finding new, better, more interesting, more fruitful ways of speaking" (360). Moreover, edification is essentially a *hermeneutic* activity; in its new mode:

> The attempt to edify (ourselves or others) may consist in the hermeneutic activity of making connections between our own culture and some exotic culture or historical period, or between our own discipline and another discipline which seems to pursue incommensurable aims. . . . It may . . . consist in the "poetic" activity of thinking up . . . new aims, new words, or new disciplines, followed by, so to speak, the inverse of hermeneutics: the attempt to reinterpret our familiar surroundings in the unfamiliar terms of our new inventions. In either case, the activity is . . . edifying without being constructive. (360)

One can detect here a certain "continental" drift but still within Rorty's constant of philosophy as linguistic activity. In the sense and to the extent that Rorty has seriously absorbed the lessons of continental hermeneutics, the notion that he has turned toward continental philosophy is only partly true. But insofar as hermeneutics is itself interpreted as linguistic activity, having to do with a kind of Rortean language game in which new vocabularies are made, discovered, or whatever, the hermeneutics takes its place within his overall analytically conceived notion of vocabularies and their unconscious assumptions. In this respect *Mirror* is simultaneously analytic and continental, while, equally simultaneously, it shifts the divisions of philosophy such that the high-level difference between foundational and edifying philosophies cuts across both analytic and continental philosophies.

Although I must return to the notion of horizontalization which plays such a crucial role in edifying philosophy, it may be of interest to see how the Rortean scheme might divide the field of phenomenological philosophies. (I note that Rorty effectively ignores phenomenology as such, and in the few instances where it appears at all it is identified, particularly in Husserl, as belonging to the foundational enterprise.) Is all phenomenology foundational? Or are there foundational and edifying phenomenologies?

Is Phenomenology Edifying?

At first reading Husserl would indeed seem to be a prime candidate for a foundational philosopher. The architectonic of his overall work clearly has that structure. He is avowedly a *transcendental* thinker, placing himself in that respect in the Descartes-Kant traditions. He "founds" his architecture upon the ground of transcendental (inter-) subjectivity. He claims phenomenology as a "rigorous science," a new science, thus not unlike positivism, that incorporates his version of a science model into his method. Moreover, most of his critics and scholarly interpreters have taken him that way.

Yet there are elements within his philosophy which are implicitly antifoundational. Interestingly, these are the devices of method which make phenomenology function as a *horizontalization* of phenomena. These are sometimes hard to detect because they must be isolated from within the apparent structure or foundationalist architectonic which scaffolds Husserl's edifice.

In this respect, the strategy of Husserl's *Cartesian Meditations* is peculiarly instructive. Husserl simply accepts the structure or framework of Descartes's project, but then, step by step, he replaces each element with a radically different result. For example, on the surface it seems that Husserl's and Descartes's foundations are the same: the *ego cogito*. But in the end, they are not. Descartes's ego is (1) the self-enclosed subject, (2) worldless except by inference or "geometric method," (3) a subject without object, and so on. Husserl's transformations replace each of these elements: (1) the (phenomenologically present) world is equiprimordial with and strictly correlated with the ego; (2) the ego, thereby, is *not* self-enclosed, and in fact is reached only by way of the world; and (3) there is no subject without an object, nor object without a subject. In short, the whole building has been replaced, and the scaffolding alone gives the semblance of a Modern philosophy.

Another, more Rortean way of phrasing this is to say that what remains after Husserl's *deconstruction* of Descartes is a new vocabulary. It is the vocabulary of the correlation of noema and noesis, of I and World, of *correlations apriori*. Moreover, if anything is "given" in Husserl, it is what is always "given" in the Rortean scheme, *some vocabulary*. Then, within this vocabulary, there are grammars of movement about how one may go in one or the other direction. I shall contend below that these hermeneutic rules at the core are the *variational methods* which derive from phenomenology. But in both senses, what remains of the Modern project is scaffolding—the problem is that Husserl was always proud of his scaffolding! This is evidenced by the vast amount of his publications which had to do with describing it in the multiple ways he did (how many reductions are there? how many ways of getting to the ego? to the phenomenological world? etc.).

It has always been my contention—admittedly disputed by many literal-minded Husserlians—that Husserl's method was heuristic. Over and over again, he adopted the terminology and the structures of Modern philosophy in both its Cartesian and Kantian forms, but in each case he reworked elements and structure, such that they no longer meant what they originally did. In his last work, *The Crisis,* he noted that what he called the "phenomenological attitude" was not, as often earlier described, a device of method, but a permanent acquisition of the philosopher. But, as I shall contend below, the result of this shift of perspective—even in Husserl—is one which is fundamentally nonfoundational.

What is hard to decide concerning Husserl himself is how much of the radicality implied in his work was discerned by him. There is no doubt that he was wedded to his terminology of "transcendental idealism," even if transcendental meant for him something radically different than in the Modern traditions and even if idealism also was intended to be different from all other idealisms. But there is little doubt that the two founders of variant phenomenologies both rejected transcendentalism (and at least by implication, foundationalism) and saw some of the more radical implications of Husserl's methodology.

Merleau-Ponty's existential phenomenology early claimed that the implication of phenomenology was not transcendental, with all the hubris of a total and self-contained system, but existential. Moreover, the late Merleau-Ponty reworks the I-World vocabulary of Husserl into an ongoing set of interrogations as in *The Visible and the Invisible.* No foundational standpoint is possible, but the polymorphy of the intertwining with its open-ended implications replace the Husserlian scaffold entirely. Similarly, Heidegger's *hermeneutic* phenomenology, although retaining a vestigial foundationalism in *Being and Time,* moves as Rorty himself has seen and appreciated in a nonfoundational direction. The expositional debate in this case revolves around Heidegger's gradual dropping of phenomenological terminology, the ambiguity as to whether what he *does* remains essentially phenomenological, and whether or not the moves into his later terminology arose through the very implications of the earlier, more explicit phenomenology.[11] Thus, if phenomenology can still be used to characterize its existential and hermeneutic versions, in its later phase it becomes ever more explicitly nonfoundational.

This excursus, however, is historical and interpretive. What is needed here to expose the edifying and nonfoundational aspects of a phenomenology is to show from its very core what motivates this possibility. Is phenomenology edifying?

To accomplish this move I shall take two seemingly contrary steps. First, I shall try to show that the edifying or hermeneutic thrust of phenomenology can be found in or to arise from essentially *Husserlian* notions (while not denying that nonfoundationalism becomes more explicit in the post-Husserlians). And, second, I shall focus not just upon the concepts and explicit claims about phenomenology, but also upon philosophical *praxis.* What dissolves the apparent contrariness between the "antique" and contemporary situations itself arises from the tradition of philosophical practice. In all of this I admit that not all *scholars* of the tradition would agree with me—but I suspect most practitioners would.

The scaffolding of which Husserl was so proud, here interpreted as a set of hermeneutic rules for proceeding, included an emphasis upon experience and evidence. Experience must be actual or fulfillable; evidence is *intuitive* (that which in fulfillment is present). What most standard interpreters have taken this to mean is that whatever is intuitively *given* not only provides some kind of foundation but belongs to the myth of the given (and Sellers was one of these interpreters). But this interpretation misses entirely the role of evidence and intuition as hermeneutic rule *to discover something else.*

For even within the heart of Husserl's explicit set of procedures, he follows what, to my mind, was a mistaken heuristic which confuses issues. The surface or explicit steps, when put into *practice,* reveal *not* the above kind of given, but something quite contrary. *For phenomenology, intuitions are constituted, not given.* Only already constituted intuitions are "given" *within an already sedimented context.* When taken as "evidence" the evidence is strictly indexical (thus hermeneutic.) What the scaffolding allows one to get at is the relationality between experience and contexts or fields. All experience is *context-relative.* Here, at a most basic level, is a first clue to phenomenological *horizontalization.*

A second "device of method," as scaffolding, is the shift from "natural" to "phenomenological" attitudes. This deliberate shift, however, is not a shift on similar levels. The phenomenological attitude is *the access to context relativity.* That is why it must become permanent—it is the now attained vocabulary wherein both the new ways of saying can be undertaken, and the inverse hermeneutic of reinterpreting the world can be performed.

This is to say, that once the structure or the field of possible contexts is open, the phenomenological attitude provides the way to explore possibilities which are its field. In the edifying sense, this means the exploration is one which seeks *to find what intuitions can be constituted.* And it is here that variational method emerges as the central driving engine of an edifying phenomenology.

Before following that implication, I would like to take note of one fundamental difference between Rorty's "hermeneutic" and a phenomenological one. Phenomenology has never been simply a linguistic philosophy, although there are strong variants, with Husserl and Merleau-Ponty as "perceptualists," and Heidegger and Gadamer more "linguistic." In all cases what counts as language is always experiential and, even better, perceptual. In phenomenology it would better be termed a language-perception pairing. Even Dasein is concretely bodily-spatial, and *Being and Time,* rather than merely talking about how one is to perform a phenomenology, undertakes it in relation to human spatio-temporality. But it must be understood that perception here means the perception of phenomenology ("lived body," "lifeworld," "time consciousness," etc.) and not that of Cartesian or Modern, neo-Cartesian physicalism.

In fact, the language-perception link in phenomenology is also tied to both context relativity and to variational praxis. For example, if intuitions are constituted, not simply given, then the task for variational explorations will be to find out in what situations, contexts, cultures, times, "x" intuition can or will occur. This is an actual hermeneutic investigative practice. Husserl preferred fantasy or imaginative variations, and through his midcareer he simply accepted the (false) empiricist assumption that imaginative variations could simply substitute for any other kind, particularly the perceptual kind (imagination duplicates perception). This practice, modeled upon mathematical procedures but also a favored shortcut for abstract and writing-room-bound philosophers, was thought to be sufficient to yield the invariants of the field of possibilities.

The post-Husserlians challenged this limitation upon variational investigation and developed other practices. Merleau-Ponty (with a different cognate disciplinary background, psychology) noted that imaginative variations could not be substituted for perceptual ones and developed this inquiry most thoroughly in *Phenomenology of Perception.* And in a more historical vein, Heidegger took the same tack with respect to his variant *epochs* of Being. Here were historical—and in the case of his "Conversation with a Japanese"—cultural variants. In each case the phenomenology involved, particularly as a practice, uncovered or hermeneutically exposed how the "intuitions" are constituted by the context. For Merleau-Ponty the context is motile bodily position with its interaction with the environment; for Heidegger it is the constellation of historical beliefs which sediment and account for some (then or now) current state of affairs.

Interestingly, this insight and practice derived directly from Husserlian phenomenology, and, in spite of explicit rejections of "phenomenology," continues brilliantly as praxis in both Derrida and Foucault! For example, in an earlier piece, I have shown how Derrida is doing something of a standard phenomenology of reading in his play upon margins and the like:

> Take a text: If one views a text (perceptually) it usually appears first as a writing that is centered on the page, surrounded by margins; but the focal center is clearly the bulk of what is written. Then, if one reads the text, what usually emerges as focal is what the text is about, however complex that may be, as indeed any text usually is. What does Derrida do with a text? Posed in the way I have indicated, he immediately decenters what seems to be focal and immediate. His focus is radically shifted to titles, signatures, margins, borders, divisions, etc. In short, he draws our attention to features that are there, but are usually taken at most as background, secondary, or unimportant features.
>
> In a sense this is a highly "phenomenological" technique. For example, in an analysis of perception, phenomenologists like to point out that while what stands out (figures) are usually most obvious because they are the referenda of our usual perceptions, all figures take their position

upon a background that is equally present and that constitutes the field of perceivability. In short, this move "decenters" focal perception so as to attend to taken-for-granted but important fringe features. Similarly, to point out that all perceptions include not only manifest surfaces, but latent "backsides," is to "decenter" at least the usual interpretations *of* perception. I am suggesting that his device—perhaps taken to Nietzschean excess—is a familiar ploy of Derrida. Indeed, one can see, once the operation is known, how to follow along with such deconstructions. (Is there a Derrida text that addresses itself to the empty background of the page? If not, there ought to be.)[12]

Foucault, too, continues the praxis of some distinctly phenomenological habits even while linking phenomenology with Husserl and opposing it. His unmentioned teacher, Merleau-Ponty,[13] remains his subterranean mentor. Foucault does histories of perception, as in the *Birth of the Clinic.* That is to say, he traces the radically different ways things are seen in correlation to the different practices of an epoch. (Foucault has a miniversion of Heidegger's epochs of Being, but Foucault's are smaller, more discrete, more rapid in change.) This praxis which continues the development of contexts of language-perception is perhaps most dramatic in *The Order of Things.* Not only is his outline a subtle response to Merleau-Ponty (who claimed there could be language about language, but not painting about painting; *The Order of Things* begins with Velázquez's *Las Meninas,* a painting about painting). The intricate pairing of experience in language-perception is precisely Foucault's forte, who may have adopted unconsciously the phenomenological vocabulary, but who *does* what I would term a kind of subterranean edifying phenomenology.

If, in Husserl's case, phenomenological edification is implicit, and if in the post-Husserlians the scaffolding and transcendentalism places existential and hermeneutic phenomenology on at least a nonfoundationalist trajectory, and if what unites this development is a certain praxis which may be either implicit or explicit, then, once having taken Rorty's shift seriously, the question can become more explicitly that of the possibility of an edifying phenomenology.

There may be a quibble here: why call it phenomenology? Heidegger ceased to use the term. Derrida and Foucault, by linking it to the foundationalist, transcendental enterprise, reject it. But if the trajectory I have outlined holds and praxis underlies what I would term an extension of variational method is the case, then there is at least more continuity than is usually allowed. Perhaps what is suggested is a *new* version of phenomenology, an edifying phenomenology which has at its core precisely that hermeneutic and inverse hermeneutic performance which freely explores what Rorty calls the exotic (histories, cultures, disciplines, etc.).

A (New) Phenomenology Which Edifies

What I have been suggesting is that phenomenology in the late twentieth century—whether it is called that or not—has had a more and more nonfoundationalist trajectory. Both the explicit transcendentalism of Husserl and the vestigial foundationalism of an early Heidegger have given way to the now-dominant strains of hermeneutics and poststructuralist enterprises of the present. Perhaps out of some unsuspected conservativism, but more likely out of a philosophical preference of actional (as opposed to epistemological) analyses, I have chosen to retain the ancestral name.

The same applies to the occasions represented here. A collection by its very nature is not a systematic or accumulative development. Rather, it presents themes, examples, applications which are united only by the unconscious vocabulary of a style—but which better would be seen as vectors along a trajectory. Husserl taught us *how* to do phenomenology. But once learned, the scaffolding which allowed one to get at the edifice is seen to be secondary. Different values about what is central emerge.

Phenomenological praxis, I would contend, revolves around an active variational inquiry. Variational inquiry may be imaginative, perceptual, historical-cultural, or interdisciplinary. It thus looks a bit like what Rorty calls hermeneutic or edifying philosophy. But also because variational inquiry is linked to a sense of experience with its language-perception pairing, there is a certain perspective upon things. It is a perspective which links a sense of *position* (from where are things seen? the vestiges of noesis) *referred* to a *context* or field (what and how do the things, old or new, appear?). But it is not tied to any preferred foundation. To be sure, if variations are perceptual, there is the privilege of the lived or motile body, but if they are intersubjective, imaginative, or whatever else, that focus is displaced.

Here, I attempt to follow the trajectory of a nonfoundationalist phenomenology. I take this to be following the *consequences* of phenomenology. The thematic occurrences of plays upon Gestalts, the development of cultural-historical variants upon the perceptions of technology, the cases of context relative phenomena are all examples along this line. Yet, while such a phenomenology may be nonfoundational and hermeneutic in at least Rorty's sense, it also may give the appearance of having a vestige of the previous past.

Phenomenology, even if nonfoundationalist, remains *structural.* But its structuralism is of an odd sort. Within its chosen field of investigations—contexts of possibility which constitute possible experiences of the language-perception type—its still essentially investigative thrust is one which discovers (vestiges of truth seeking) a multiplicity of structures. I return once more to the notion of multistability to make the point.

Structures discovered are not all of one type. In some of my examples of visual multistability, I would contend that the structure of possibility is linear and arbitrary (contrary to Merleau-Ponty). But such a possibility structure in no way exhausts the possibilities of others. Were one to move from the abstract, two-dimensional drawing examples, to concrete, three-dimensional objects in the normal Earthbound context, one might discover a structure of *graded* possibilities.

Vary the Washington Monument: its current stability is upright.

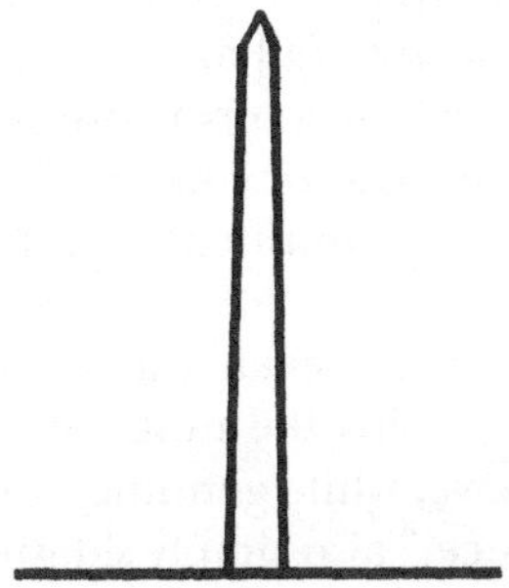

This actual stability, without changing its architecture, would be even more stable, hence a graded possibility, were it to lie on its side on the level ground:

And under some temporary conditions it *might* even be (barely) stable upside down and perfectly balanced:

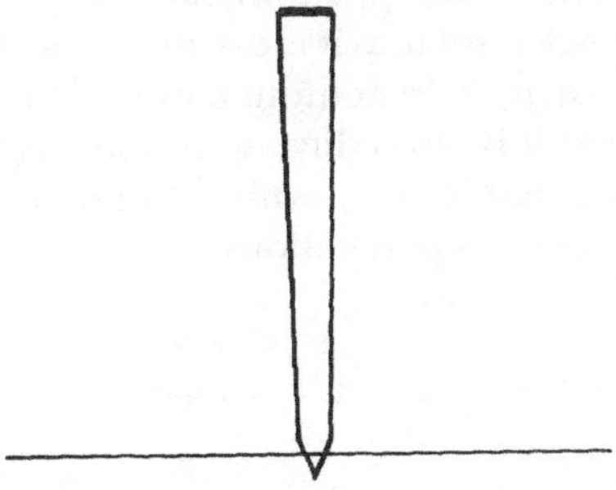

But this last possibility is clearly gradedly weaker than either of the first two, while other possible positions are so weak as to be impossible (without changing the structure of the monument itself):

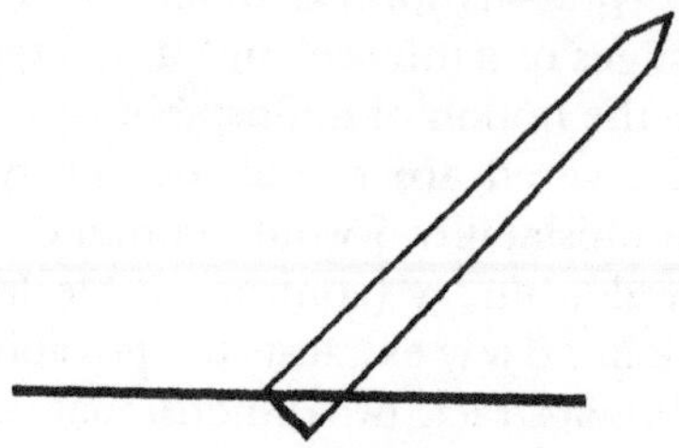

Still other types of structural stability could be hierarchical, serial, independent-dependent, and so on. If this is vestigial "foundationalism," it is both oddly so, since the investigation and horizontalization of the field of structure is neither selective (all are context relative) nor reductive (there is no "best" or "only" structure). But it does come short of the one aspect of Rorty's conversations which place him much closer to the newer French versions of continental philosophy than to the older phenomenological ones.

By removing both truth seeking and referentiality entirely from edifying philosophy, Rorty joins the ranks of the poststructuralists and deconstructionists who have, while genuinely creating a new type of historical and cultural "science," also simply sidestepped the possibilities of what I prefer to call a *noematic* science. (The natural sciences, interestingly enough, come closer to this sense of phenomenological praxis—of the investigation of possible structures—than the previous human sciences.) This is in no way to deny what Kuhn and the new philosophy of science have discovered, that the practice of the sciences falls under essentially hermeneutic interpretations, that science itself includes an inevitably hermeneutic dimension, and so forth. But the need for the "outward" look, the noematic reference, is implicitly the retention of the language-*perception* pairing found within phenomenology.

Phenomenology insofar as it is essentially hermeneutic is edifying in Rorty's sense. That is, it is nonfoundational in its newer and post-Husserlian forms. But it is also edifying in a stronger sense for it is not without "edifice," *structure.* Finally, while phenomenology does have its vocabulary, it also retains its perceptions.

10

Why Not Science Critics?

The idea for my title was suggested quite a few years ago by Langdon Winner. Langdon had sent me a copy of a collection of his essays to read and respond to which eventually became *The Whale and the Reactor.* And, although his topic was philosophy of technology and his experience was what many of us in technology studies felt at the time, the point Langdon made applies equally well to science, or even better, to what is now often called *technoscience:*

> [This] project . . . is a work of criticism, a fact that some readers will find troubling. If, in contrast, this were literary criticism, everyone would immediately understand that the underlying aim is positive. A critic of literature examines a text, analyzing its features, evaluating its qualities, seeking a deeper appreciation that might be useful to other readers of the same text. In a similar way, critics of music, theater and the arts have a valuable, well-established role, serving as a bridge between artists and audiences. Alas, the criticism of [technoscience] is not welcomed in the same manner. Writers who venture beyond the most ordinary conceptions of tools and uses, writers who investigate ways in which technical forms are implicated in the basic patterns and problems of our culture are met with the charge that they are merely "antitechnology" [or "antiscience"] or "blaming [technoscience]." All who have stepped forward as critics in this field—Lewis Mumford, Paul Goodman, Jacques Ellul, Ivan Illich, and others—have been tarred with the same brush, an expression of a desire to stop the dialogue rather than expand it.[1]

The contrast between art and literary criticism and what I shall call "technoscience criticism" is marked. Few would call art or literary critics "antiart" or "antiliterature" in the working out, however critically, of their products. And while it may indeed be true that given works of art or given texts are excoriated, demeaned, or severely dealt with, one does not usually think of the critic as generically "antiart" or "antiliterature."

Rather, it is precisely because the critic is passionate about his or her subject matter that he or she becomes a "critic." That is simply not the case with "science" or technoscience criticism.

In part this is because art and literary criticism is *institutionalized.* It is so much a part of the artistic and literary tradition that critics meet, publish, and talk in the same contexts as the artists and writers. And, contrarily, there simply is no such forum within science or technology. The critic—as I shall show—is either regarded as an outsider, or if the criticism arises from the inside, is soon made to be a quasi outsider. Why is this the case?

We are now at the juncture where I may announce the theses which I wish to argue. First, I am obviously holding that I think something like "technoscience criticism" *ought* to be part of the social discourse concerning technoscience and that this role ought to be a recognized and legitimated role. And, in a related fashion, while I am holding that the technoscience community resists precisely this role and bears the self-serving primary responsibility for closing off the critical dialogue, I am also implicitly holding that "technoscience criticism," while often occurring, has not taken the place it could occupy in the places where we should expect it to occur. This contention comes from my own experience in furthering the growth of North American "philosophy of technology"—which does frequently offer technology criticism and frequently gets for its efforts an "antitechnology" label—and from my more recent experience and work in the philosophy of science, which until recently has almost strenuously *avoided* anything which could be called "science criticism" except in the narrowest of conceptual senses. I shall later examine a few instances of technoscience criticism which have and do occur, but which are, at best, limitedly successful.

1. Barriers to Technoscience Criticism

The most obvious barrier to the formation of an "institutionalized" technoscience criticism lies in the role of late Modern technoscience itself. Technoscience, as institution, began in early Modernity by casting itself as the "other" of religion. Its mythologies, drawn from classical pre-Christian and often materialist (Democritean/Epicurean) sources; its anti-authoritarianism, including the Galilean claim to have exceeded the Scriptures and church fathers' insights, founded in the new sighting possible through his telescope; and the much stronger later antireligiosity of the Enlightenment, which cast religion as "superstition" and science as "rationality"—all led to the Modernist substitution of what I am calling *technoscience* for religion.

In the process, science—whether advertently or inadvertently—itself took on a quasi-theological characteristic. To be critical of the new "true faith" was to be, in effect, "heretical," now called "irrational." Functionally speaking, this resistance to criticism serves to keep the critics externally located, as "others." And while none of this is news, it maintains itself within the institutional characteristics of technoscience's own belief structure.

The success of this science-religion inversion is instanced in the child's textbook version of Columbus's voyage. We grew up believing that while Columbus knew or at least believed the world to be spherical (hence he was a rational, scientifically informed navigator), his crew believed the world to be flat (hence they were religious and superstitious) and if he went too far they would fall off the end of the earth-ocean. This myth about the fifteenth century, as Valerie Flint showed in her *The Imaginative Landscape of Christopher Columbus,*[2] was itself an early-twentieth-century *invention.* Indeed, the time of invention, now incorporated into our pre–late-twentieth-century deconstruction of Columbian history, was when the Scopes trial was underway. What Flint showed was that even the most moderately informed individual of the fifteenth century believed the earth to be spherical, with the exception of a very small, obscure group of "flat-earth" sectarians. The early-twentieth-century inventors of the rational-scientific versus superstitious-religious binary simply elevated the texts of the flat-earthers beyond proportion and claimed this was a widespread belief. But it was an apparently successful polemic which did get institutionalized into our science-dominated education insofar as many children still believe in the invented story. This is but one example of a dominant mythology which still functions.

However, the science-religion inversion is too general to account for the resistance to an institutionalized technoscience criticism. Instead, I wish to turn our attention to two features of technoscience which are both more deeply embedded in technoscience praxis and which relate more closely to the art-literature criticism analogy.

The first relates to *science texts.* Bruno Latour's *Science in Action* argues that science-as-institution has successfully created a social form which contains its own form of critique into carefully constructed modes of contestation. Science-as-institution incorporates structured *trials of strength* through which controversies are settled without damage to the basic institution itself.

In a strategy derived from both phenomenology and deconstruction, Latour deliberately *inverts* what we usually take to be the scientific self-interpretation—the interpretation which usually begins with an inquiry into Nature and ends with a well-formulated "law" or theorum concerning a natural phenomenon finally promulgated in a text—by beginning with results, a "text" or scientific article, and working reconstructively backwards to appeals to Nature.

Typically, scientific "texts" or literature appear as articles in scientific journals. I cannot trace out here the extreme complexity of the construction of such articles, which in Latour's interpretation are carefully crafted results of trials of strength, but everyone is familiar with the fact that virtually all such articles are (1) multi-authored, (2) written in a deliberately anonymous or authorless style, and (3) couched in both quantitative and visualized chart forms. Here, already, is a very "unliterary" form.[3]

Latour asks: Who reads such texts? And how are such texts—for my purposes here—"criticized" or challenged? First, most people do not read such "texts" at all! Rather, the readers are already usually members of a select community. Indeed, the *technical opacity* of such texts is part of the form of the text itself. The text "puts off" any ordinary reading. The text does not invite one in unless the reader is already "an expert" in that style of reading.

But, then assuming that one knows how to "read" such a "text," what are the possible outcomes? Latour cites three: first, one simply "goes along" with the text. One accepts, believes, and, if in the field, quickly incorporates the findings which then become part of the larger system of science-as-institution. Or, if you wish to challenge or criticize the "text," you find that it can't be (often) done through textual criticism per se, but you have to go to an entirely different level—you have to go to the laboratory which produces the conditions for the text. Were there to be a count-requirement for literature, it would be one whch required us to return to the Agora to refute Plato! Here is Latour's version of what happens:

> The peculiarity of the scientific literature is now clear: the only three possible readings all lead to the demise of the text. If you give up, the text does not count and might as well not have been written at all. If you go along, you believe it so much that it is quickly abstracted, abridged, stylized and sinks into tacit practice. Lastly, if you work through the authors' trials, you quit the text and enter the laboratory. Thus the scientific text is chasing its readers away whether or not it is successful. Made for attack and defense, it is no more a place for a leisurely stay than a bastion or bunker. This makes it quite different from the reading of the Bible, Stendhal or the poems of T. S. Eliot. (ibid.)

I would like to go on into the context of the *laboratory* and on into the appeal to *Nature*, which is where the leaving of the text leads in technoscience, but I cannot. I can only say that to challenge the text at the level of the laboratory ultimately calls for one to construct a counter-laboratory if the challenge is to be carried through. This is a thoroughly technologically embodied process which implies money, gangs of operators, and even an educational process. But it also leaves the potential critic in a very unusual and uncomfortable position.

This points to the second inherent problem for the development of technoscience criticism. And that is the dimension of knowledge-power or, better put, knowledge-*expertise* which functions within late modern technoscience. In this context I shall turn to Raphael Sassower for illumination from his *Knowledge without Expertise.* In this book, Sassower turns to an interesting early Modern example which occurred in the British Association for the Advancement of Science. The internal BAAS debate was whether to keep or expel Section F from its ranks and thereby effectively "defrock" *economics* from the status of being a "science." This historical example shows well how knowledge relates to power, but in the particular form of "expertise" whereby only "experts" are empowered to make decisions. Generalizing on this modern form of technocracy, Sassower notes:

> If accepted, the myth [of expertise] has an immediate pragmatic consequence since it suggests that only experts can and should make decisions about their speciality, and that only experts in the same field may judge each other's decisions. What about the non-experts? They seem unqualified to be external reviewers of the decisions of experts, for they do not possess the specialized knowledge that qualifies experts to make certainty claims. In this sense, then the myth of expertise guarantees, . . . that experts judge other experts and that experts are shielded and even insulated from public reproach.[4]

I probably need not remind many here how the myth of expertise operates as two-edged sword in so many academic contexts: in philosophy, for example, shall the dominant philosophical traditions (by number still analytic philosophers) be the sole arbiters of what counts as philosophy? Or does the counter-ploy of counter-expertise—only continental philosophers should judge continental results—come into play? But in the realm of technoscience criticism the usual role that expertise plays relates to the claim on the part of science-as-institution that only the scientifically informed may be certified as critics.

I want to enter two examples of criticism in action here, to show how the critic is made "other," external to institutionalized technoscience.

The first instance is autobiographical, and as critic (a philosopher) I was already identified as an external critic.

The occasion was an interdisciplinary panel of scientists, convened to debate the issue then facing Long Islanders about the Shoreham nuclear plant. The plant had just gone low-level operational, prior to the approval of an evacuation plan (imagine evacuating Long Island!) in the case of a nuclear disaster. One of the panelists was Max Dresden, an acerbic and outspoken physicist who had been associated with the Manhattan Project. His presentation turned out to be a defense of expertise and a diatribe against even allowing public discussion of expert conclusions. He contended (this was before Chernobyl) that nuclear energy was the cleanest, safest, and ultimately the cheapest source of electric power and that it was "irrational" to oppose—out of ignorance—the opening of the Shoreham plant.

During the discussion, I entered the fray, at first provoking Max to reiterate in even stronger terms his defense of expertise—he now claimed that no one should be allowed to *vote* on issues of such technical complexity. So I asked him whether the Shoreham debate was a "scientific" or a "political" debate, and he eagerly admitted, albeit ruefully, that it was "political." I then asked him if he were expert at politics, and he huffed and said no, to which I replied, then, according to your expertise argument you ought to have nothing to say about politics, but merely leave it up to the political process to have its day. (Of course, the Shoreham plant has been decommissioned!) The next day at the faculty club, Max came over and attacked me as loudly as possible, saying he wished the entire Philosophy Department could be dismantled given its "antiscientific" tendencies! I probably need not further explicate how this illustrates the externalization of criticism from within an institutionalized "myth of expertise."

The second instance is one which begins with the critic as insider, a "whistle-blower" example: I suspect everyone remembers the news coverage of the 1991 "Gulf War." It was a trial run on one of our "Star Wars" developments, the antimissile missile, the "Patriot." The news broadcasts showed over and over again the presumed "interceptions" and claimed hits up to 95 percent effectiveness. If, then, you followed the more critical analyses to come, you will probably recall that there was an admission that effectiveness or "hits" declined to about 24 percent. Part of this admission was due to the early-on analysis performed by Theodore Postal, a ballistics expert and MIT scientist who took news videotapes used to make the hit claims and subjected them to magnified, enhanced, and computer image techniques, which on closer inspection showed that claimed hits were not hits at all. Eventually, he concluded *that there may not have been a single, verifiable hit* made by a Patriot! Needless to say, this claim was not appreciated by Raytheon, the manufacturer of the missile, nor by his colleague, Shaoul Ezekiel, who had advised Raytheon, and eventually not even by MIT itself, which got caught in the crossfire of claims and anticlaims.

The battle turned nasty: Raytheon implied that Postal had actually doctored the tapes, but they later reduced this to the claim, suggested by Ezekiel, that the grain structure and imaging of videotapes was simply too gross to make the conclusions drawn. The battle continues to this day, particularly between Postal and Ezekiel concerning ethical conduct, with MIT trying to shy away because of the large amounts it gets annually from Raytheon.[5]

Nor is this some isolated instance. In a study of the "costs of whistle-blowing," *Science* reports that more than two-thirds of whistle-blowers (within science as an institution) experience negative effects ranging from "ostracism" through "pressure to drop allegations," to the actual nonrenewals or loss of jobs.[6] The long-drawn-out David Baltimore case is another of these scenarios, in which the whistle-blower—not the offender who faked the notebooks—was fired. The insider critic is isolated and, if possible, often separated and thus made into an outsider or "other."

While the above scenario would not be much different for business corporations, neither would we be surprised about this ostracization from the corporate sector. But because of the popular image of science as more like a church in its claims about critical concern for truth, this may come as a surprise, although not for those of us close enough to realize that science-as-institution is today much more like the corporate world than it is a church!

2. Science Criticism

The implicit trajectory clearly shows that there is a role for science criticism. This is not to say there is none—quite to the contrary, the examples cited show that there is both external and internal criticism which does take place, regardless of costs. But, equally, the role of criticism is not one which parallels the role of the art or literary critic, nor is science criticism validated in the same way. Rarely, unless the criticism is so extreme as to provoke public outrage, is an art or literary critic fired, ostracized, or threatened. Moreover, the sector in which one would expect such an institutionalized criticism to originate, namely the philosophy of science, has also not performed this task adequately.

I do not have space here to trace out the reasons for this lacuna, other than to suggest that the heretofore dominant traditions of the philosophy of science (derived from positivist and analytic traditions, more recently from pragmatic analytic traditions) did indeed take the passionate view of their subject matter which the presumed literary critic takes toward literature, but the result was not criticism so much as an attempt to *justify* and even to *imitate* science, in short, to make philosophy more "like" science.

And when philosophy of science did become normative, it did so in the name of an idealized rationalistic conceptualism. Early positivist attempts to isolate the pure logical form of science and then normatively judge science resulted in the laughable results which proclaimed such sciences as geology "unscientific" or relegated most of the biological sciences into a kind of "softness" akin to sociology.

Two other areas of philosophy came a little closer to establishing science criticism: I refer to the various types of "applied ethics" domains which arose most prominently in the medical contexts, and aspects of the philosophy of technology, in which case the degree of tellingness of critique remains marked by the accolade "antitechnology" as noted by Winner above. But each is at best a partial success. Applied medical ethics has matured and has become partially institutionalized inside medical schools and hospitals, and it does perform evaluative and reflective exercises. Philosophers of technology, on the other hand, have been prone to be far too generic in criticism, both by reifying technologies as Technology with the capital "T," and by making too sweeping claims about "alienation," the subsumption of "Nature" to "Technology," and so forth. Throughout, both external and internal critics remain to be taken as "others." So do we end with failure? Or with the impossibility of *science criticism* in anything like an analogy to art and literary criticism?

3. What Would a Science Critic Look Like?

Given this state of affairs, it now behooves me to make some projections. If I am calling for science critics, what would they look like? What would they do? And where would they be?

Continuing to pursue the art-literature analogy, I would say the science critic would have to be a well informed, indeed much better than simply well-informed *amateur,* in its sense as a "lover" of the subject matter, and yet not the total insider. Increasingly some philosophers of science have caught this: Ian Hacking, in his work on science instruments, particularly the microscope, has called for philosophers to "go native" in some degree. And I also agree with him that this is more a matter of learning *science practice* with its forms of tacit and operational know-how than it is a matter of conceptual analysis.

Second, again like the art or literary critic, it is probably equally important that the amateur *not be* a fully practicing artist or literary writer. Just as we are probably worst at our own self-criticism, that move away from self-identity is needed to position the critical stance. Something broader, something more interdisciplinary, something more "distant" is needed for criticism.

So far, so much like art and literary criticism. But I think techno-science is in many ways a special case which calls for more than occurs in art and literary criticism. For one thing, art and literary criticism still follows the "authorship paradigms" of its subject matters. Critics remain, like writers and artists, individuals who both do and like to sign their own names, to be personally responsible, and thus bear both the praise and blame for results. Science is not like that: its style is anonymous, impersonal, and, above all, corporate or intersubjective. But underneath, even the process of discovery is an intersubjective and increasingly multiple perspectived process. I should like to suggest that the critic, the philosopher, must more thoroughly enter into this process.

To do this, one aspect is collaboration. At Stony Brook, we have one example of such a collaboration in the results from our Logic Lab and the co-authored work of Patrick Grim and Gary Mar on computer-modeled philosophical problems. Their research—although not an example of science criticism in any direct sense—has reached as high as notice from *Scientific American*[7] and is noted for the innovations in fuzzy logics and game theory based on cooperative models. We simply must "gang up," and produce through cooperative and intersubjective work our critical results.

Another aspect, related to both the "going native" and the collaborative result, is that the science critic must get in on the origins, rather than the results, of the technoscientific process. I have long argued that one flaw in applied ethics fields is that they are like the ambulance corps which attends the battlefield—they fix up the wounded but do not either prevent the battle or ameliorate its consequences. Only when the critic is, in this metaphor, present at the strategy planning of the generals can the critic hope to affect the outcomes.

Here examples of critical participation at foundational stages are even rarer and harder to broach since the exclusionary forces within science-as-institution mitigate against such presence. But I can cite examples (I cite a recently discovered such example in my *Philosophy of*

Technology,)[8] mostly some I have discovered in the last several years in trips to northern European technical universities.

In Scandanavian and Dutch technical universities, philosophers have found themselves within *research teams*, and while sometimes assigned the evaluation and consideration of ethical and social outcomes in assessment contexts, sometimes they find other skills are called for. Increasingly, I have found myself drawn into these contexts by being asked to review and respond to research design.

I will end with one autobiographical example. In Denmark a team of researchers are dealing with a certain problem of medical crises which occur in operating rooms. Picture the patient, unconscious and anesthesized, but hooked up to an array of machines around the room which give readings of vital signs. Dials, audible alarms, oscilloscopes—all are part of the hermeneutic display to be "read" by the practitioners. In turn, each device is programmed to go off at a preset level. During the operation, however, most alarms that go off, experienced physicians have learned, are "false alarms," and here one reaches a certain critical juncture. If one ignores the alarm and it is "genuine," clearly there is a danger to the patient; yet, on the other hand, if the alarm is "false" and one stops to fix the situation, other dangers occur.

The dilemma discovered was that the most experienced physicians were more likely to ignore the alarms—sometimes with disastrous results—more often than less experienced physicians. So the problem became one of how does one determine a "real" from a "false" crisis? Moreover, this was determined to be an explicitly "hermeneutic" problem, a matter of right reading. But, like science, the result is not the fictive world, but the "nature" or patient beyond the instrumental texts. What does the critic do, or what can the critic do? It is here that I place him or her to enact *science criticism*.

PART 4

EXPANDING HERMENEUTICS

This final section of the book is a minimonograph on its overall aim: to expand the role of hermeneutics into technoscience. Bits of the suggested program found in some of the chapters preceding part 4 are now synthesized and made into a coherent trajectory. The reigning, positivist image of science has often enough been proclaimed "dead," but a reframing for the understanding of science has not always been undertaken. Here I try to show how hermeneutics, interpretive activity, occurs within the sciences—but not just any interpretive activity, and not what is more likely to be taken as hermeneutic within literary and humanistic contexts. Rather, I am demonstrating in the analysis of science praxis the unique way science has been able to create a *visualist hermeneutics* which, while it functions in a way analogous to the much earlier invention of writing, functions through the various dimensions of the visualization of things.

This hermeneutic, not unlike all forms of writing, is technologically embodied in the instrumentation of contemporary science, but focally in its development of visual machines or *imaging technologies.*

I conclude speculatively, but within the context of even more radical technologies. The simulative arts and the virtual reality technologies of the present, I argue, point to possibilities which, while retaining the bodily-perceptual activity which phenomenologists favor for constructing knowledge, point to a possible different kind of interpretive activity. Can there be a "whole-body hermeneutics"? To get beyond visualism is the question.

11

The Field Is Clear

The essays which make up the first three parts of this book now are behind us. Part 4, "Expanding Hermeneutics," is the last part of this hybrid production. It is here that I want to follow a program which takes hermeneutics into the sciences and to show how science can do a "hermeneutics of things" by turning them into scientific objects. I shall undertake this project in three steps. First, one preliminary is to clear the field for hermeneutics within science, to reconverge what began to diverge with early modernity. One has to see that this reconvergence has already begun and to take note that the *field is already clear.* The clearing process has, in fact, had several waves in the dissolution of what I have called the Hermeneutics-Positivist Binary (H-P Binary)—changes in the philosophy of science, the emergence of a new sociology of science, and the rise of feminist critics. Then I note that there is already a minoritarian, but dynamic, group of hermeneutic philosophers of technoscience who have made initial hermeneutic forays into science.

Once the field is seen to be clear, I then follow a "weak program" of identifying hermeneutic dimensions implicit within current science praxis, followed by a "strong program" which examines the cutting edges of science's knowledge constitution in a hermeneutic way. The expansion of a "thingly hermeneutic" concludes with an outline of a reconstructed understanding of science along hermeneutic lines.

1. Field Clearing

Part 1 of this book covered the location and development of hermeneutics, first from its own history, and then in an initial foray into the philosophies both of technology and of science, but which for purposes here I combine as technoscience. Chapter 1 traced hermeneutics in European thought up to its "graft"—as Ricoeur called it—to phenomenology.

Then in chapter 3, I first began to explore the role of hermeneutics in the philosophy of technology. There I referred to the shock I experienced in the first meeting of the International Society for Hermeneutics and Science (ISHS), a meeting which was a critical factor in stimulating the present program.

As noted, the "conservative" hermeneuts at that first conference remained largely Diltheyan; that is, hermeneutic methods were seen as appropriate only to the *Geisteswissenschaften.* As Karl Otto Apel put it, one can do a hermeneutic history of science, or a hermeneutic sociology of science, but not a hermeneutics of either science or its objects. This view, at the time, was a shock to me, since I did not regard the Diltheyan split between the human sciences and the natural sciences to be as clear in the late twentieth century as it might have been at its beginning. Moreover, while distinctly minoritarian, there already existed, particularly in North American contexts, the beginnings of a clearly "hermeneutic" philosophy of science which were far more radical than that. And, even prior to this postclassical hermeneutics, the very graft of hermeneutics to phenomenology should have undercut the Diltheyan program, since the "ontologizing" of hermeneutics placed *both* the *Geisteswissenschaften* and the *Naturwissenschaften* under "ontic" restraints. Put simply, if rudely, it seemed to me that the Diltheyan cast to this strand of European hermeneutics was *distinctly old fashioned and not up to date.*

A. Deconstructing the hermeneutics-positivist binary

To show this, I need to make, initially, a short survey of what has happened vis-à-vis hermeneutics and the various contemporary interpretations of the sciences. I begin with the deconstruction of the H-P Binary. This binary has two sides: the first shows that several versions of positivism became what amounted to the "standard view" of the construal of science, and the second was the internal acceptance of aspects of positivism even, or perhaps especially, by classical phenomenological and later the dominant European phenomenological-hermeneutic philosophers (P-H traditions).

In its broadest outline, as the sciences, particularly post-Enlightenment, attained success after success, the rhetoric of science became increasingly and stridently antireligious and antimetaphysical. This rhetorical thrust remains very much a constant of textbook science even today. The master narrative is one which hails the rise of *reason* and rationality over superstition and religious belief (which belongs to a dark past). In its most excessive form, this rhetoric becomes as guilty as its predecessors of distortion and mythic construction.

One of my own hobby interests has always been the history of sailing and of navigation. And for Europeans, Christopher Columbus stands forth as one of the innovators of transoceanic navigation. I have followed that history for a long time, including the multiple inventions of "Columbus," who basically was forgotten or certainly not forefronted until three centuries after his set of voyages to North America (1492–1503). His first recognized "centennial" was in 1792, not coincidentally as part of the celebration of the nationhood of the United States after the establishment of its Constitution (1789). Of heroic stature then, through 1892 Columbus becomes the "discoverer of America" and "Admiral of the Ocean Seas"—even though he never knew it and died believing he had reached the islands off Japan!

What I did not realize until recently was the extent to which the master narrative about him was "constructed" quite deliberately as part of the rhetorical program referred to above. Just prior to the quincentennial (1992), Jeffrey Burton Russell published his *Inventing the Flat Earth.*[1] Russell shows, with standard—not deconstructionist—historical analysis that not only is the master narrative about Columbus false, but that it was quite deliberately constructed as part of the movement to counteract certain religious revivals and particularly fundamentalism in the late nineteenth and early twentieth centuries (including the Scopes trial controversy then raging).

Russell demonstrates that no educated persons in Columbus's time believed in the flat earth, that the fears of the "superstitious" sailors about falling off the end of the Earth were unfounded and unlikely to have existed at all, and that with minoritarian exceptions (including the geographer chosen by Columbus), most geographers had a fairly good idea about the size of the Earth and of the possibility of a westward route. I had already been quite familiar with the double log kept by Columbus on his voyage, the public log which was open to his crew and the secret log he kept for himself which showed larger daily runs than the open log showed. He obviously wanted to keep the greater distances from his crew. Their fear, however, was *not* about falling off the Earth, but of going so far that, given prevailing trade winds, the ship could never return. This was a *rational* fear, in that sailing ships of the time could not sail into the wind (as current sailing vessels can—to within at least 35 degrees), and no European (at least, since the Vikings) knew of the world patterns of the wind beyond coastal navigation. There was no empirical reason to believe that westward winds would reverse.

Columbus's theory—which turned out to be the heart of his navigational innovation—was that the winds had to be circular, with southern trade winds matched by northern returning winds. This guess was correct, and he was able to return to Spain precisely by turning north and capturing these heretofore unknown winds. The narrative about superstitious sailors, however, played into the rhetorical need to make Columbus a prefigure of scientific rationality and enlightenment (which, in most respects, he was not). My shock, however, related to the deliberateness with which what was basically a known falsehood could be incorporated into this construction.

I cite this story because it very closely corresponds to the double march of the H-P Binary. The positivist interpretation of science—which, I hold, became "the standard view"—held for almost precisely the same time span as the modern division of hermeneutics *from* science. Its first expression occurs with Auguste Comte (1798–1857). Comte is writing at the same time as Friedrich Schleiermacher (1768–1834), who actually exacerbates the H-P Binary by turning hermeneutics into a psychologized "philosophical anthropology."

Comte's version of positivism founds much of the master narrative rhetoric referred to above: Comte borrows a "three-age" scheme from earlier thinkers (including Vico) and argues that science arises and becomes established once the Age of Religion and the Age of Metaphysics are overcome. There follows the Age of (positive) Science. Three-age scenarios are not new in European thought, but they usually contain not terribly hidden normative patterns. Modern three-age scenarios invert the ancient ones which are usually ages of deteriorization (Gold/Silver/Iron and Clay), and in modernity become progressivist. Each subsequent age supercedes, and thus devalues, the preceding age. Moreover, by overcoming, the present (scientific) age is more strongly distinguished from its predecessors. The first positivism is thus a kind of progressivist, evolutionary positivism. Seemingly implied here—if this is the origin of the H-P Binary—is a possible association of hermeneutics with religion, metaphysics, or both, and of positivism with science.

But that is too strong an association. Admittedly, the earlier use of hermeneutics in "language"-related senses—texts, sacred texts, writing—did place hermeneutics close to the humanistic and linguistic disciplines. Yet hermeneutics as it turned modern, and that only late when compared to the sciences, had also become secularized and more general. The division between hermeneutics and science, however, remained and was exacerbated by the second wave of positivism.

The twentieth-century successors to earlier positivism are, of course, Logical Positivism and Logical Empiricism. Taking Comtean science, already separated off from religion and metaphysics (traditional philosophies as systems), and interpreting the sciences as *hypothetical-deductive* forms of explanation, second-stage positivism simply cedes all higher rationality to this process. Science not only becomes more autonomous, but becomes construed as a kind of logical "machine" for theory production (and empirical verification). The Logical Positivist theory goes so far as to eliminate any sense of truth from other than logical formulations (tautologies) or empirically verifiable (i.e., scientific) statements.

If hermeneutics is the other side of a binary, this leaves it simply "outside" science—and, in effect, that is precisely where Wilhelm Dilthey placed it, giving it a degree of autonomy, but also divorcing it from explanation or the ability to understand the "nature" of the natural sciences. My point here is not to reenter the debates which gave rise to this binary, but to point to the historical and functional result: this separation of hermeneutic methods from science (1) cedes science and its construal to positivism and (2) prevents the analysis or appreciation of the deep hermeneutic elements to be found in actual science praxis. It was this virtually mutually accepted H-P Binary which I found to be operational within the understanding of science among some of the founders of ISHS.

B. The "new" philosophies of science

The H-P Binary, however, does not occupy the place it once did, even in the philosophies of science. Indeed, positivistic programs in the philosophy of science have been declared moribund now for several decades and by most within the field itself. Steve Fuller, in a genre review of some eighty-seven books in the philosophy of science published between 1977 and 1988, claims that among these, "only [one] widely discussed book to extend the logical positivist program since 1977 [was published]."[2]

This is an old story, and I shall not detail it except in outline. By the late fifties the strong version of positivism was already under attack by Karl Popper and Imre Lakatos, who effectively reduced verification to a combination of falsification plus consensual research programs. Paul Feyerabend, perhaps the most radical of the iconoclasts, although active in this time, followed Thomas Kuhn with *Against Method.*[3] But it was Kuhn who both won the antipositivist battle and ended up the new "reframer" of science construal. *The Structure of Scientific Revolutions*[4] remains the most cited history and philosophy of science book in the twentieth century. The most important indicator, however, is the use of "Kuhnian" language in the self-interpretation of scientists which can be found in virtually every issue of *Science, Scientific American,* and other widely read science magazines. (I shall be using very recent issues of these same science publications in illustrating science's "hermeneutics.")

I shall not trace the initial battles, the rhetorical charges against Kuhn (to which he often overcompensated), but refer to the general trends which emerged after him. If Logical Positivism and Logical Empiricism largely faded, what emerged was a field of much more modest approaches to the sciences in which science gets interpreted as a fallibilist, finite, problem-oriented, and often regional set of inquiries into different subject matters. Usually still claimed to be the most reasonable and self-correcting of the extant forms of rationality, nevertheless philosophies of science now dominant could be called *"pragmatic."* Philosophers as diverse as Ian Hacking, Hilary Putnam, Richard Rorty, Larry Laudan, Thomas Nickles, insofar as they address science, seem to fit this category. This process, begun in the late fifties, continues to the present. (I shall refer to more specifics in the section on emergent hermeneutic philosophies of science.)

C. The sociologies of science

A second set of field-clearers, from a different perspective, began to gain ground, largely in the seventies on: the so-called social constructionist sociologists of science. Again, I shall not explore here the often bitter and intense battles between the philosophers of science and the sociologists of science, other than to remark that the rhetorics look much like those earlier directed at Kuhn.

The new "post-Mertonian" sociologists of science, in the simplest perspective, apply the methods of the various social sciences to *science-as-institution,* thus making of science a culture. That is, the anthropologists, sociologists, and even quasi-philosophical social theorists look at science-as-institution as another form of social organization with rules, practices, and behaviors to be examined. This seems a perfectly valid thing to do, but it does shift focus away from what science *claims* it does and thus opens the way to a misunderstanding concerning science claims—which, in turn, have led to the debates about the new sociologies.

But in the development of the various versions of social constructionism, there is also a shift of focus as to what is most interesting about science. In stark contrast to the "theory machine" of earlier positivism, the sociology of science looks at daily practice, usually at what goes on in experimental and laboratory science. Early and strongly discussed work centered on the "strong program" of Bloor et al., the look at laboratory life by Steve Woolgar and Bruno Latour and by Karin Knorr-Cetina, and the examination of how scientific "products" are "constructed" as in Andrew Pickering's *Constructing Quarks.*[5]

The uproar which has followed is one by which the sociology of science practitioners' approach was seen (by both some scientists and philosophers of science) to reduce science knowledge products to just other social products. This placed the sociologists of science firmly, even more radically, in the "antirealism" camp of some philosophers of science.

Again, I shall not follow out the details, but note in passing that, first, there is some indirect connection here with the P-H traditions among these thinkers. Pickering, in particular, draws from the "social construction of reality" tradition of Berger and Luckmann, who in turn draw heavily from Husserl and Schutz, so there is a quasi-phenomenological background lurking here. Similarly, the popular and highly visible work of Bruno Latour clearly draws from the continental traditions, and the shadows of both Derrida and Foucault are not hard to detect. Second, whatever the outcome, the intrusions of the sociology of science have at least shifted the parameters of where the reconstrual of the sciences can take place. Even the most conservative philosophers of science now acknowledge there to be a "social dimension" to science,[6] and the shift of focus to praxis—particularly in the more specific looks at experiments and laboratories, and even extending to the politics of science—is evident.[7] If a hard-core logical empiricist is hard to find today, so is it hard to find a science interpreter who denies the social embeddedness of science-as-institution.

D. Feminist critics of science

The third wave of field-clearers has been the feminist critics of science. Here, the eighties mark a high-water trend which continues. The feminist critics do not follow a single trend but have subtraditions which are neo-Marxist (standpoint theorists, among whom Sandra Harding stands out), neo-Freudian or developmental (Evelyn Fox-Keller and others), and post-modernist perspectivalists (Donna Haraway).[8] But there remains strong agreement among most feminist authors that science is a thoroughly socially and culturally embedded phenomenon—*which is also marked by the signs of patriarchalism* not unlike other Euro-American or industrially developed societal structures.

At the very least, the demographic gender maldistribution, in which the "harder" the science, the more male-dominated, and in which the deep embeddedness of male-dominant rhetoric which paces so much of the history of science—Nature as a female to be domesticated or manipulated—remain echoed even among recent Nobel laureate speeches, seems to be well established. The defense posture, in which "science" is made into a kind of ideal type, divorced from the actualities of practice, simply gets shown to be a defensive response which has no empirical actuality. Thus, again, as institution, science gets perceived as a thoroughly social, cultural, and political process (as well as remaining a producer of knowledge, although no longer seen as a simply idealistic, rational, and selfless search.)

The purpose of this survey, admittedly sketchy, is to show that the field is clear and its boundaries shifted. Science, as a special activity, does take its place *within* the larger, and more complex, lifeworld. It begins to look a good deal like a process which follows particular modifications within the lifeworld, as Husserl had claimed and foreseen. And, while the above three movements within the philosophies of science, the social sciences, and feminist critics do not necessarily address the hermeneutic side of the H-P Binary, they do show that the acceptance of the binary particularly as regards the construal of science is no longer to be taken for granted.

E. Hermeneutic beginnings in the philosophy of science

The clearing already undertaken primarily addresses the positivist side of the binary. I now turn to its hermeneutic counterpart. I have previously noted that a minoritarian, although active, hermeneutic tradition concerning the philosophy of science has emerged. In chapter 4, I introduced the work of Hubert Dreyfus and Patrick Heelan, both with original hermeneutic programs related to the sciences. Here I should also acknowledge two earlier hermeneutic philosophers, Joseph Kockelmans and Ted Kisiel, who have both introduced the role of Heidegger vis-à-vis science, and who have done substantial scholarly work relating the hermeneutic tradition to science. More recently, a philosophically inclined physicist, Martin Eger, and my colleague Robert Crease have also entered the scene. Eger has done interesting work relating and comparing a hermeneutic to a social constructionist approach, arguing for the greater appropriateness of the former approach. Crease, using a theatrical production metaphor, has shown how scientific objects are "prepared" in experimental contexts, a theme I shall take up below from a different persective.[9]

Nor should the philosophers who come from the Anglo-American traditions (analytic), but who have delved into hermeneutics, be ignored. Here I list Robert Ackermann, Ian Hacking, Richard Rorty, and with respect to an indirectly hermeneutic approach to experiment, Peter Galison. I have addressed in much greater depth these developments in my *Instrumental Realism.*[10]

Given the thrust of this program, however, the two thinkers who are most relevant to the "praxis-perception" approach I am taking are Joseph Rouse through *Knowledge and Power* and Bruno Latour in *Science in Action.*[11] Rouse shows how the hermeneutic approach to science becomes relevant within the new philosophies of science, and Latour develops a somewhat postmodernist approach to hermeneuticizing science.

Rouse, in *Knowledge and Power,* with the subtitle *Towards a Political Philosophy of Science,* follows Heidegger and Foucault in his resituation of science. One must note that there has been a trajectory since Kuhn to move science away from the earlier predilection with "theory" as a central preoccupation toward a preoccupation much more with praxis. Kuhn's addition of "history" did enrich the image of science in action, but *Structure* remained, nevertheless, predominantly a history of *theory.* The sociologists of science, shifting to the laboratory as the site of science in action, make this trajectory even more concretist, and Rouse belongs to this shift as well.

Rouse sees a convergence of pragmatist-oriented philosophers of science with the growing minority of *hermeneutic* trends in interpreting science:

> Much of [the] sudden upsurge of interest is due to the recognition that hermeneutics and pragmatism reinforce each other and even converge in some important ways. Thus it should not be surprising that the philosophers who have most extensively discussed the importance of hermeneutics for the philosophy of science—Rorty, Habermas, Bernstein, and Hesse—have also been prominently associated with the revival of pragmatism.[12]

Rouse reinforces the point I have made about the H-P Binary, by noting what others today, such as Charles Taylor, have claimed: "Old-guard Diltheyans, their shoulders hunched from years of long resistance against the encroaching pressure of positivist natural science, suddenly pitch forward on their faces as all opposition ceases to the reign of universal hermeneutics" (47).

But what is this reign of "universal hermeneutics" which now begins to reinterpret not simply the history of science, but science praxis itself? Rouse helpfully, following Dreyfus, notes that there are two forms of "universal hermeneutics." He distinguishes between a transitional version, the *hermeneutics of translation* (or *theoretical holism* in Dreyfus), which meshes nicely with postanalytic forms of the philosophy of science, and a more radical *hermeneutics of practice,* which follows more Euro-American traditions.

I will not follow out the full explication of Rouse's interpretation, other than to note that the hermeneutics of translation or "theoretical holism" opens the way into the analytic pragmatism of today's postanalytic philosophers (Davidson, Putnam, Rorty, Laudan, et al.) by taking acts of translation into epistemology. This version of hermeneutics retains its *explicitly linguistic heritage* into recent epistemologies of belief, conceptuality, and truth theory. Such a hermeneutics relates to accounts of how we scientifically can come to understand the workings of the world.

However positive this "analytic" hermeneutics may be, Rouse prefers the more radical Heideggerian *hermeneutics of practice* as the form of universal hermeneutics for science itself: "[This] version of universal hermeneutics has its origins in Heidegger and the later Wittgenstein. Interpretation is taken to be the working out of the possibilities open within a situation, rather than the translation of theories or beliefs" (48). Clearly, this is a more "ontological" trajectory than theoretical holism. It is not, however, a phenomenologized perceptual, nor instrumentally embodied trajectory.

I shall not follow Rouse farther here, since it is enough to note that the reclamation of science by hermeneutic interpretations is not something idiosyncratic, but a well-recognized trend, particularly in North American circles. It is the trajectory which brings into convergence hermeneutic and pragmatic interests in the philosophy of science itself.

From Rouse, I turn to Bruno Latour, sociologist-philosopher, who works both in France and the United States. His *Science in Action* is the work most relevant here. Latour, like Rouse, draws much from Foucault and, perhaps indirectly, from Derrida. His approach, at first, seems close to many in the new sociology of science. He proposes as a methodological focus:

- To put aside truth claims (within science) and take account of operations—thus we have an indirectly praxis orientation, and
- By inverting the processes ordinarily thought, to characterize natural sciences, so that he then can note its workings.

That is, instead of beginning with the interrogation of "Nature," Latour begins with the *texts* of science—scientific articles—and moves backwards and downwards from these to the layers of references embedded in social practices.

The moment which interests me here is the move from texts (science articles) to the *laboratory*. In Latour's context the question is one of how one determines the "truth" of claims made in science. Texts themselves are insufficient—one may read scientific articles—but textual criticism, even logical analysis, and one might as well say also any of the classical hermeneutics of understanding, all turn out to be insufficient to the test of claimed "truth" of the texts. Instead, one must move outside the text, into the place where the tests are performed—*the laboratory.*

It is with regard to the laboratory that Latour plays a rather neat "hermeneutic trick." The laboratory, according to Latour, is not only the place where scientists do their work, it is the place *where inscriptions are produced* (pace Derrida!). For Latour, *instruments* are, in effect, inscription-producing devices. The instrument is what lies behind and beyond the text: "This move through the looking glass of the paper allows me to define an *instrument* . . . I will call an instrument (or *inscription device*) any set-up, no matter what its size, nature and cost, that provides a visual display of any sort in a scientific text."[13]

Let us from the beginning take note of the steps being taken in Latour's hermeneuticization of the laboratory. (1) The text is never autonomous but refers beyond itself to the work which produces and lies beneath the text (the text is designed to efface itself). (2) That reference is to the work which produces the claim of the text, to the laboratory where an *instrument* is set up to produce an *inscription* or *visual display:* "The instrument, whatever its nature, is what leads you from the paper to what supports the paper, from the many resources mobilized in the text to the many more resources mobilised to create the visual displays of the texts" (68). Then, lest we miss an insight previously elaborated from Heidegger's tool analysis and on through my own development of a phenomenology of instruments,[14] the instrument, while producing the visual display, *is not itself "visible" or forefronted:*

> What is behind a scientific text? Inscriptions. How are these inscriptions obtained? By setting up instruments. This other world just beneath the text is invisible as long as there is no controversy. A picture of moon valleys and mountains is presented to us as if we could see them directly. The telescope that makes them visible is invisible and so are the fierce controversies that Galileo had to wage centuries ago to produce an image of the Moon. (69)

Thus, if Latour is right, the instrument is already *a hermeneutic device.* And, equally, hermeneutic practice lies in the very heart of the laboratory. The laboratory has now become something like "science's scriptorium."

We have now advanced several steps beyond classical hermeneutics. We are now, at least, inside science practice, within the working laboratory, and here hermeneutics occurs. And, insofar as Latour has turned the laboratory into science's scriptorium, there is one more side to this process. The laboratory not only prepares inscriptions—but it is the place, the site, where things—scientific objects—are prepared or *made readable.* This is where a peculiarly scientific hermeneutics begins.

12

Scientific Visualism

If you are convinced by the narrative history just traced, then the expansion of hermeneutics, already convergent toward the sciences, should be ready to *enter* into the realm of the sciences and their "thingly" concentrations. Premodern hermeneutics naively believed that Nature contained the "writing" placed there by God in the creation. But we cannot return to that context or time. I shall use here a different and constructed metaphor to replace the ancient one. The postmodern hermeneutics of things must find ways to give *voices* to the things, to let them *speak from themselves*. The source of this metaphor is phenomenological and draws from my own research history. For both Husserl and Merleau-Ponty, voiced language is a *bodily* and a fully perceptual activity. It is that materiality of perceptual activity which I seek in this thingly hermeneutic.

Nearly three decades ago, in my own research history, I began investigating *auditory* perception, which plays such important human roles in speech and listening (and music), roles which I thought the tradition had often neglected and which thus sometimes led to the impoverishment of philosophical richness in understanding human being-in-the-world.[1] One of the lessons which emerged from these investigations was that so many things which are simply seen or viewed appear *silent*. Yet everything potentially *has a voice* under one of two conditions: first, seemingly silent things can be given voices—that has frequently been the musician's gift to our experiences. Even stones, struck, "speak" forth. Percussion is one way of giving voices to things.

The voice which is given to things, or elicited bodily from things, however, is very *complex*. First, if the thing is struck on the model of musical percussion, the voice is not single but is a duet. The sound produced is both the voice of the thing struck and the voice of the striking instrument. Substituting a bell for the stone example, a bell struck with a wooden mallet produces a different "duet" of sound than one struck with a brass mallet, or again with a rubber mallet. Here there is some implicit *science* as well. The relativistic and quantum sciences of the late twentieth century are self-consciously aware of this instrument-object interface. Even to measure the temperature of a pot of water blends together the eventual temperatures of the water and of the measuring thermometer. As we shall see later, this becomes one practical reason for introducing what I call "instrumental phenomenological variations" into our letting things speak, that is, the use of multivariant instrumental measurements.

The second condition is one which can appear only at the microlevel. Often a thing may already be "sounding" below or above the levels of our "earhole" perceptions (my substitute here for the common use of the "eyeball" observational). The radiations of wave phenomena—microwaves above or below earhole perceivability—may again become presentable to our hearing *if mediated by the proper instrument.* Thus, for a second time, the material intervention of a material artifact, a *technology,* enters into the conditions for giving things a voice, or allowing the thing to "speak" for itself.

As I am writing this, "Pathfinder" and its little robot, "Sojourner" (named for Sojourner Truth), are exploring rocks on Mars. The examples from metaphorical auditory experience are obviously apt for this set of events. The Sojourner, equipped with an alpha proton X-ray spectrometer, must approach each rock carefully, touch it, and bombard the rock with X-rays, thereby probing the ions in the rock by radiative "percussion"—the "giving voice" metaphor, here at the microlevel, is appropriate to the high-tech, engineering-embodied space science of the present. The ions activated "bespeak" the chemical composition of the rock.

What, however, is "hermeneutic" in all this? To answer that is the real task of this chapter. Following in a slightly different way the suggestions of Rouse previously noted, I want to differentiate between what I shall call a "weak" and a "strong" hermeneutic program with respect to science. The "weak" program is an attempt to reconstruct accounts of science praxis, showing the implicit *hermeneutic practices* already at play within science. This amounts to a claim that in one of its knowledge-producing dimensions, science is already a hermeneutic practice. Here the task is to show that various interpretive activities within science practice are already hermeneutic in form.

The "strong" program is potentially more normative. It will be an attempt to push, positively, certain P-H practices by way of suggestion and adaptation toward science practice. Thus, I will be outlining a more aggressive hermeneuticizing of science, although based upon forefront research fields as now emerging in late Modern science.

1. Modifying the P-H Tradition

The first step in expanding hermeneutics into the thingly sciences is to *modify* the phenomenological-hermeneutic tradition itself, clearing it of some of its own last prejudices about science, but from within some of its own potential insights. The thrust of this modification is one which reconstructs the understanding of science and the lifeworld. As I argued in several of the previous chapters, most explicitly in chapter 4, classical phenomenology—particularly with Husserl of the *Crisis* and "The Origin of Geometry,"[2] and with Merleau-Ponty in *Phenomenology of Perception*[3]—tended to interpret science as separated off from the basic perceptual-bodily activity of the lifeworld.

This separation, however, comes about in part because the construal of science in early phenomenology was in keeping with the P-H Binary I have alluded to, the binary which tended to view science as a propositional and theory-biased special activity, rather than the institutionalized, embodied, and *technologically mediated* science which late Modern philosophers have more often taken it to be. In short, I am arguing that this science, *technoscience,* has never been separated off from the lifeworld but is a unique Modern, now-postmodern, activity which produces its knowledge through its very technological embodiment.[4]

Of course, this means that the P-H tradition must accept within itself the mediated forms of intentionality which come through technologically mediated experiences, alongside and with all other bodily perceptual activities. Science, in this view, is one highly developed, complex, and skilled mode of mediated praxis, but in principle it is not out of step with Heidegger's "hammer" or Merleau-Ponty's "feather."

To accomplish this shift within P-H understanding of science, I shall revert to a Galilean parable—a revisitation to a fictional, phenomenologized Galileo as Godfather of modern science. The historical Galileo, while "metaphysically" proclaiming a reduced and abstract world of inertial motion of material objects, in practice employed an array of instruments through which his early science made its discoveries. While not the inventor of the compound lens telescope, Galileo developed and perfected over a hundred of these instruments. He became convinced that telescopically mediated perception was "better" than eyeball perception, and one of his arguments involved evidence that a certain "halo" around celestial objects could be seen with his telescopes which could not be detected by naked eye.[5] The irony is, of course, that this effect was what we know as an "instrumental artifact," an effect of the technology, not of the referent object.

We also know, historically, that Galilco proclaimed that "any man" could see what neither the ancient philosophers nor the church fathers could see—but, only under the condition that Galileo "teach" the process of telescopic seeing.[6] The initial perceptual ambiguities of planetary features and satellites were, again, associated with the primitive technology of Galilean optics. This stage of early, often hard to resolve object referents, only later resolved by better instrumentation, has long been a feature of the history of science.[7] The handheld telescopes of Galileo were difficult to use and, particularly, to "fix" upon the referent objects. Beginning astronomers today can note this same problem set when viewing through handheld binoculars, for example.

Third, the historical Galileo clearly had his fascination centered upon the celestial objects, on what was "out there," and gave these observations his primary attention. This, again, is a typical feature found associated with the uses of new technologies. Its echo today can be found in the hype accorded to all sorts of technologies, particularly computers. There is a tendency to overestimate what can be gained with many novel technologies.

But, now, I want to reinvent Galileo, in a more thoroughly phenomenological style. He has, for some time now, been tinkering with his optical instruments and has already noted the surface features of the biggest nighttime object, the Moon. When ancient theory called for the Moon to be smooth surfaced, purely circular, and of a celestial perfection not possible upon Earth, the very first sightings had to have been dramatic. Even Galileo's instruments showed—immediately and dramatically—craters, valleys, mountains, plains, shadowed and lighted irregularly with respect to the phase of the Moon vis-à-vis its light source, the Sun. One cannot blame him for the initial enthusiasm that this first discovery of an "instrumentally mediated *realism*" which the telescope delivered. Nor could one avoid the later sense which would maintain itself into the contemporary world, of "effects which will simply not go away."[8]

We now, however, need to *phenomenologize* this fictional Galileo. He begins his description of the phenomenon, at first noematically, that is, with a concern for the "out there" object:

Spatiality has just undergone a set of dramatic changes; suddenly the Moon has mountains, craters, and so on, which mean that what was previously more "distant" is now "closer." But what makes it closer, and what changes occur? First, the "closer" Moon (through the telescope) has now displaced its previous context. It no longer occupies its relatively located and smaller appearance within the overarching heavens. In relation to its previous field, it has radically changed. Magnified, the Moon is "closer."

Did the Moon simply change its distance? Phenomenologically, every noematic shift is correlated with a noetic shift, that is, with a shift of the positioned perspective which is refracted from every visual observation *back to the position of the embodied observer.* This new, instrumental distance is thus a difference in distance between my body and the Moon (within the mediation). Indeed, it is equivalent to say that the Moon is now "closer" to me, or I am "closer" to the Moon—it is the *relational, or relativistic, distance* which has changed.

This phenomenon, historically, became known as "apparent distance." Why apparent? One possible answer is that the "real" distance between Galileo and the Moon did not change: it remained 240,000 miles distant whether or not he saw it through the telescope. This, of course, either privileges the measurable distance between the actual body of the observer and the referent object (thus, implicitly, does this mean that the eyeball observation is privileged as well?) or is constructed by some variation upon Cartesian-Newtonian space which presumes—not, as it claims, a nonpositional measurement—a measuring stance from an ideal or god's eye perspective which simultaneously sees both Galileo and the Moon.

I have described this version of "apparent" versus "real" distance in this way to show that there is lurking here a *variation* between eyeball and instrument-mediated perceptions. If the reflexivity of intentionality is maintained, it is the entire Gestalt which changes between direct and mediated perceptions. The relative distance, telescopically, is reduced. And the bringing near of the Moon occurs within the new context.

Phenomenologically, the relational distance is the *intentionality distance* which must include both referent object and perceiving, perspectival "lived" body, but not in the same way as in Cartesian-Newtonian frames. This phenomenological measurement must be reflexive and must utilize means which determine the (apparent) distance from within the correlation. What must be avoided is the ideal observer or god's eye simultaneous sight. Instead, were I to invent a measuring technology, it could be something like a radar probe in which the known speed of the signal is timed against its return. This space-time method is reflexive and avoids the Cartesian-Newtonian implied external perspective.

The new experience also implicates the very sense of one's body as well. In the beginner's experience, this is easily detectable when first trying to "fix" the celestial object. One can notice the "wavering" of the object, but this wavering is simultaneously the wavering of my holding of the instrument. The Moon's magnified character is simultaneously the *reflexive magnification of my bodily motion.* Both the Moon and I have changed, and both areas are possible areas for further and deeper investigation. If I didn't know the Moon had mountains (before instrumental mediation), neither did I know that I constantly had micromotions to my bodily position when actively viewing something other than myself. But both noematic and noetic transformations are equally detectable in this new experience.

But, again within the context of changed Gestalts, there is a detectable difference in one's sense of body. The now suddenly micromotional body experience also conveys a sense of "irreality" when compared to my usual, already familiar actional sense of motion upon the Earth. A small indication of this occurs whenever anyone gets a new prescription for eyeglasses: one has to relearn, in however minuscule ways, the distancing between the ground and walking, and so on. Once learned, the "irreality" either is diminished or virtually disappears as the instrument is properly "embodied" into the new (now normalized) experience. Our phenomenological Galileo already knows how to use the telescope and can teach others to use it by applying his own experience.

The historical Galileo's interest, however, remained outwardly, externally oriented. And that interest can become—became—a *technological trajectory* which could be refined and followed. To enhance "fixing" the celestial object, first tripods, later more complicated machinery built into the now optical system, which includes motorized devices to keep the telescope fixed upon the section of the Moon being studied, was the trajectory historically followed.

These had to entail growing knowledge of Earthly and heavenly motions, to bring to a halt for observational purposes what was always in relative and now magnified motion into the technological development. Second, these developments which help to "fix" or stabilize observations without bodily training are also developments which can and do "distance" the previously direct bodily activities from the original contexts. The actual—relativistic—bodily seeing is replaced by the technoconstruction which allows the vision to be mechanically stabilized. The Greek preference for the eternally fixed remains within the trajectory being followed. Machines can "embody" metaphysics.

A trajectory is, however, a *choice*. In the original Galilean situation, there was multistability, and with it a number of possible choices. For example, had Galileo been fascinated with the revelation of his bodily micromotion discovered through use of the telescope, he might have taken a different direction. And later investigators did follow that trajectory. Contemporary empirical psychology employs a range of instruments which focus precisely upon micromotions of the human body. Microtiming of response time, minuscule eye motion, reaction time as a factor lying between autonomic and voluntary responses, and so on all were implicit from the instrumental beginnings, but historically came much later in the histories of the sciences.

This is to say that the historical Galileo *chose* certain, rather than other, interests, followed one, rather than multiple, instrumental trajectories, and in the process, insofar as he was followed by a community of like-minded fellows, opened *a* pathway for modern science which early favored astronomy, physics, and geometry.

Before leaving our reinvented Galileo, I want to follow one more line of phenomenological inquiry with respect to the *lifeworld*. Wanting to observe several of the dramatic comets which became visible in the last few years, I purchased a high-quality reflecting telescope, which my family, visitors, and I have enjoyed in the clear summer nights in Vermont. The Moon, the easiest target in the sky, is also the easiest to use to demonstrate "instrumental realism." A puzzle, however, has begun to emerge: when we look at the Moon with the naked eye, it always appears to have light and dark areas (the old "Man in the Moon" phenomenon), such that my son and I began to wonder how the ancients could ever have thought the Moon to have a pure, mirror-like, featureless surface. We simply *cannot see* it that way. But has our seeing already become post-telescopic seeing, and the Moon has changed, not just under the actual conditions of telescopic viewing, and now "contains" within the eyeball example the "residue" of the telescopic?[9]

One even more dramatic such effect occured in 1996. We had been viewing Jupiter, whose four largest moons are dramatically visible through the telescope. These moons move in their orbits and thus clearly and quickly change position from night to night—another easy case of instrumentally delivered "realism." Then, one clear night, driving home from a nearby concert, I happened to glance up and saw—with the naked eye—Jupiter "with horns," or a clearly visible line out to each side of the disc. And I immediately recalled that Sandra Harding refers to Dogon (African) observations in the eleventh to thirteenth centuries which refer to Jupiter in just such a way.[10] I had simply rediscovered an ancient observation but, having done so, now have a "different" Jupiter within both eyeball and telescopic vision. Like the textured Moon, Jupiter no longer can revert to simple disc. This is, in short, a lifeworld accretion which follows an irreversible direction. While there are different Gestalts for naked and mediated perceptions, there is also an interaction and overlap which through familiar embodiment shapes the contemporary texture of the lifeworld.

This revisitation of Galileo has been intended to show something concretely about a P-H-revised understanding of science within the lifeworld. From here, the expansion of hermeneutics, first as elicited from actual practices of science and then into extrapolations of a more deliberate sort, can be done.

2. The "Weak Program": Hermeneutics Implicit within Science

In its broadest (and earliest) sense, hermeneutics could be rendered "interpretive activity." The metaphors of a more *linguistic* sort have tended to dominate, from the "Book of Nature" to my own "giving voice" regarding things. Add postmodernist emphases upon "textuality," and we note that much of interpretive activity has fallen under linguistic metaphors.

This is often the case even within the sciences, particularly if linguistics includes its *semiotic* dimensions. It is apparent in a transparent way within contemporary genetic sciences, where one now has strings of "codes" which can be determined and deciphered: genes "express" themselves in a technical sense which is not semiotic, but which carries the vestige of linguistic overtones. The same may be said of computer science with its binary coding and of digital processes in which "bytes," "pixels," and other "codes" are invoked, again carrying the linguistic overtones noted. One can even add the intermixture of genetic and computer semiotics as not only roughly parallel, but probably arising from a more or less implicit *Zeitgeist* which favors this set of grand metaphors. These metaphors both are more sophisticated than the earlier dominant "mechanical" metaphors and are able to carry the greater degrees of *complexity* which contemporary science demands. The ancient mechanical clock now seems to be replaced by the digital computer as the instrument of choice for our metaphors. If this is a trace of hermeneutics in science—and I think it is—it remains an accidental, rather than a deliberately constructed, one.

The vector I have been taking, however, is a somewhat different one. I have been arguing for and demonstrating a much more *bodily* and *perceptualistic* mode of interpretive activity. This moves somewhat tangentially to the dominance of the linguistic, and substitute features from perceptual experience which I am holding carry an appropriateness for the *thingly* which I take not to be somehow reducible to the secretly "linguistic."

I am tempted here—although I realize the dangers—to note that perceivability is an action which overlaps humans and animals probably more than language activity does. Of course, qualifications must immediately be entered: our perceivability is intertwined with linguistic and cultural interconnections which are distinctly human. That is because our "worlds" are also unique insofar as we are seldom prey who are actually eaten, when compared to our cousins, both domestic and wild, which frequently find this pattern in their lifeworld. The indicators, though, become somewhat more dramatic in the cases in which we try to enter, penetrate, or translate animal "languages" or teach them variants of our own. At best, there are suggestive intimations with chimpanzees or dolphins, and the specter of Quinean intranslatability is much more problematic with these quasi-languages than in the case with natural languages among humans—even among philosophers! Yet we might suspect that perceivability—of fast-moving objects, of oncoming objects, of changes in atmosphere—contains a greater overlap between us and our cousins.

If this is so, please note that I am not therefore arguing that perceivability is a "lower" or more "primitive" action than linguistic action as such. Quite to the contrary, I am arguing parallel to many of Dreyfus's observations that perceptual-bodily activity is both basis and implicated with all *intelligent* behavior.[11] And, if my broader overlap speculation is valid, then for an interpretive activity with the thingly, we need something more than "textuality."

Much of the line I have argued with respect to the history and philosophy of science is that Modern to late Modern science is what it is because it has found ways to enhance, magnify, and *modify* its perceptions. Science, as Kuhn and others after him seem to emphasize, is a way of "seeing." Given its explicit late Modern hyper-*visualism,* this is more than mere metaphor. There remains, deep within science, a belief that seeing is believing. The question is one of how one can see. And the answer is One sees *through, with, and by means of instruments.* It is, first, this perceptualistic hermeneutics that I explore in the weak program.

A. Scientific visualism

It has frequently been noted that scientific "seeing" is highly visualistic. This is, in part, because of historical origins, again arising in early Modern times in the Renaissance. Leonardo da Vinci played an important bridge role here, with the invention of what can be called the "engineering paradigm" of vision.[12] His depictions of human anatomy, particularly those of autopsies which display musculature, organs, tendons, and the like—"exploded" to show parts and interrelationships—were identical with the same style when he depicted imagined machines in his technical diaries. In short, his was not only a way of seeing which anticipated modern anatomies (later copied and improved upon by Vesalius) and modern draftmanship, but an approach which thus visualized both exteriors and interiors (the exploded style). Leonardo was a "handcraft imagist."

The move, first to an almost exclusively visualist emphasis, and second to a kind of "analytic" depiction, was faster to occur in some sciences than in others. In astronomy, analytic drawing of telescopic sightings was accurate early on and is being rediscovered as such today. The "red spot" on Jupiter was already depicted in the seventeenth century. But here, visual observations and depictions were almost the only sensory dimension which could be utilized. Celestial phenomena were at first open only to visual inspection, at most magnified through optical instrumentation. It would be much later—the middle of the twentieth century—that astronomy would expand beyond the optical and reach beyond the Earth with instruments other than optical ones.

Medicine, by the time of Vesalius, shifted its earlier tactile and even olifactory observations in autopsy to the visualizations à la da Vincian style, but continued to use diagnostics which included palpations, oscultations, and other tactile, kinesthetic, and olfactory observations. In the medical sciences, the shift to the predominantly visual mode for analysis began much later. The invention of both photography and X-rays in the nineteenth century helped these sciences become more like their other natural science peers.

Hermeneutically, in the perceptualist style of interpretation emphasized here—the progress of "hermeneutic sensory translation devices" as they might be called—*imaging technologies* have become dominantly visualist. These devices make nonvisual sources into visual ones. This, through new visual probes of interiors, from X-rays, to MRI scans, to ultrasound (in visual form) and PET processes, has allowed medical science to deal with bodies become transparent.[13]

More abstract and semiotic-like visualizations also are part of science's sight. Graphs, oscillographic, spectrographic, and other uses of visual hermeneutic devices give Latour reason to claim that such instrumentation is simply a complex *inscription-making device* for a visualizable result. This vector toward forms of "writing" is related to, but different from, the various isomorphic depictions of imaging. I shall follow this development in more detail later.

While all this instrumentation designed to turn all phenomena into visualizable form for a "reading" illustrates what I take to be one of science's deeply entrenched "hermeneutic practices," it also poses something of a problem and a tension for a stricter phenomenological understanding of perception.

Although I shall outline a more complete notion of perception below, here I want to underline the features of perception which are the source of a possible tension with scientific "seeing" as just described. Full human perception, following Merleau-Ponty, is always *multidimensioned* and *synesthetic*. In short, we *never just see something* but always experience it within the complex of sensory fields. Thus the "reduction" of perception to a monodimension—the visual—is already an abstraction from the lived experience of active perception within a world.

Does this visualizing practice within science thus reopen the way to a division of science from the lifeworld? Does it make of science an essentially reductive practice? I shall argue against this by way of attempting to show that visualization in the scientific sense is a deeply *hermeneutic practice* which plays a special role. Latour's insight that experiments deliver *inscriptions* helps suggest the hermeneutic analogy, which works well here. Writing is language through "technology" in that written language is inscribed by some technologically embodied means. I am suggesting that the sophisticated ways in which science *visualizes* its phenomena is another mode by which understanding or intepretive activity is embodied. Whether the technologies are translation technologies (transforming nonvisual dimensions into visual ones), or more isomorphically visual from the outset, the visualization processes through technologies are science's particular hermeneutic means.

First, what are the epistemological advantages of visualization? The traditional answer, often given within science as well, is that vision is the "clearest" of the senses, that it delivers greater distinctions and clarities, and this seems to fit into the histories of perception tracing all the way back to the Greeks. But this is simply *wrong*. My own earlier researches into auditory phenomena showed that even measurable on physiological bases, hearing delivers within its dimension distinctions and clarities which equal and in some cases exceed those of visual acuity. To reach such levels of acuity, however, skilled practices must be followed. Musicians can detect minute differences in tone, microtones, or quarter tones such as are common in Indian music; those with perfect pitch abilities detect variations in tone as small as any visual distinction between colors. In the early days of auditory instruments, such as stethoscopes, or in the early use of sonar, before it became visually translated, skilled operators could detect and recognize exceedingly faint phenomena, as clearly and as distinctly as through visual operations. Even within olfactory perception, humans—admittedly much poorer than many of their animal cousins—can nevertheless detect smells when only a few molecules among millions in the gas mixture present occur in the atmosphere. In the realms of connoisseurship such as wine tasting, tea tasting, perfume smelling, and the like, specifics such as source, year, and blend—even down to individual ingredients—can be known. It is simply a cultural prejudice to hold that vision is ipso facto the "best" sense.

I argue, rather, that what gives scientific visualization an advantage are its *repeatable Gestalt features* which occur within a technologically produced visible form, and which lead to the rise and importance of *imaging* in both its ordinary visual and specific hermeneutic visual displays. And, here, a phenomenological understanding of perception can actually enhance the hermeneutic process which defines this science practice.

Let us begin with one of the simplest of these Gestalt features, the appearance of a figure against a ground. Presented with a visual display, humans can "pick out" some feature which, once chosen, is seen against the variable constant of a field or ground. It is not the "object" which presents this figure itself—rather, it is in the interaction of visual intentionality that a figure can appear against a ground.

In astronomy, for example, sighting comets is one such activity. Whether sighted with the naked eye, telescopic observation, or tertiary observations of telescopic photographs, the sighting of a comet comes about by noting the movement of a single object against a field which remains relatively more constant. Here is a determined and trained figure/ground perceptual activity. This is also an *interest-determined* figure/ground observation. While, empirically, a comet may be accidently discovered, to recognize it as a comet is to have sedimented a great deal of previous informed perception.

These phenomenological features of comet discovery stand out by noting that the very structure of figure/ground is not something simply "given" but is *constituted* by its context and field of significations. To vary our set of observables, one could have "fixed" upon any single (or small group) of stars and attended to these instead. Figures "stand out" relative to interest, attention, and even history of perceivability *which includes cultural or macroperceptual features* as well. For example, I have previously referred[14] to a famous case of figure/ground reversibility in the history of aesthetics. In certain styles of Asian painting, it is the background, the openness of space, which is the figure or intended object, whereas the almost abstract tracing of a cherry blossom or a sparrow on a branch in the foreground is now the "background" feature which makes space "stand out."

When one adds to this mix the variability and changeability of instruments or technologies, the process can rapidly change. As Kuhn has pointed out, with increased magnifications in later Modern telescopes, there was an explosion of planet discoveries due to the availability of detectable "disc size," which differentiated planets from stars much more easily.[15]

I have noted in the previous section that Latour, in effect, sees instruments as "hermeneutic devices." They are means by which *inscriptions* are produced, visualizable results. This insight meshes very nicely with a hermeneutic reconstrual of science in several ways.

If laboratories (and other controlled observational practices) are where one prepares inscriptions, they are also the place where objects are made "scientific," or, in this context, *made readable.* Things, the ultimate referential objects of science, are never just naively or simply observed or taken, they must be *prepared* or *constituted.* And, in late Modern science, this constitutive process is increasingly pervaded by technologies.

But, I shall also argue that the results are often not so much "textlike," but are more like repeatable, variable *perceptual Gestalts.* These are sometimes called "images" or even pictures, but because of the vestigial remains of modernist epistemology, I shall call them *depictions.* This occurs with increasing sophistication in the realm of *imaging technologies* which often dominate contemporary scientific hermeneutics.

To produce the best results, the now technoconstituted objects need to stand forth with the greatest possible clarity and within a context of variability and repeatability. For this to occur, the conditions of instrumental transparency need to be enhanced as well. This is to say that the instrumentation, in operation, must "withdraw" or itself become transparent so the thing may stand out (with chosen or multiple features). The means by which the depiction becomes "clear" is constituted by the "absence" or invisibility of the instrumentation.

Of course, the instrumentation can never *totally* disappear. Its "echo effect" will always remain within the mediation. The mallet (brass, wood, or rubber) makes a difference in the sound produced. In part, this becomes a reason in late Modern science for the deliberate introduction of *multivariant* instrumentation or measurements. These *instrumental phenomenological variations* as I have called them also function as a kind of multiperspectival equivalent in scientific vision (which drives it, not unlike other cultural practices, toward a more postmodern visual model).

All of this regularly occurs within science practice, and I am arguing that it functions as a kind of perceptual hermeneutics already extant in those practices. I now want to trace out a few concrete examples, focused upon roles within imaging technologies, which illustrate this hermeneutic style.

Galileo's hand-held telescopes undertook "real time" observations, with all the limitations of a small focal field, the wobbliness of manual control, and the other difficulties noted above. And, while early astronomers also developed drawings—often of quite high quality—of such phenomena as planetary satellites, the isomorphism of the observation with its imaged production remained limited.

If, on the other hand, it is the repeatability of the Gestalt phenomenon which particularly makes instrumentally produced results valuable for scientific vision, then the much later invention of *photography* can be seen as a genuine technological breakthrough. Technologies as perception-transforming devices not only magnify (and reduce) referent phenomena, but often radically change parameters either barely noted, or not noted at all.

It would be interesting to trace the development of the camera and photography with respect to the history of science. For example, as Lee Bailey has so well demonstrated, not only was the camera obscura a favorite optical device in early Modern science, but it played a deliberately modeling role in Descartes's notion of both eye and ego.[16] From the camera obscura and its variants to the genuine photograph, there is a three-century history. This history finally focused upon the *fixing* of an image. As early as 1727, a German physician, Johann Schultze, did succeed in getting images onto chalk and silver powders, but the first successfully fixed image was developed by Joseph Niepce in 1826. His successor, Louis Daguerre, is credited usually (in 1839), but Daguerre simply perfected Niepce's earlier process.[17] I shall jump immediately into the early scientific use of photography.

If the dramatic appearance of relative distance (space) was the forefront fascination with Galileo's telescope, one might by contrast note that it is the dramatic appearance of a transformation of *time* which photography brought to scientific attention. The photograph "stops time," and the technological trajectory implicitly suggested within it is the ever more precise micro-instant which can be captured. In early popular attention, the association with time stoppage often took the association between the depiction and a kind of "death" which still photography evoked. Ironically, the stilted and posed earliest photos were necessary artifacts of the state of the technology—a portrait could be obtained only with a minutes-long fixed pose, since it took that long for the light to form the negative on glass covered with the requisite chemical mixture.

Photography, however, was an immediately popular and rapidly developing new medium. And, if portraits and landscapes were early favored, a fascination with motion also occurred almost immediately. The pioneers of stop-motion photography were Eadweard Muybridge and Thomas Eakins at the end of the nineteenth century. Muybridge's studies of horses' gaits served a popular scientific interest. He showed, with both galloping horses and trotters, that all four feet left the ground, thus providing "scientific" evidence for an argument about this issue, considered settled with Muybridge's photos of 1878.[18] Insofar as this is a "new" fact (this is apparently debatable since there are some paintings which purport to show the same phenomenon), it is a discovery which is instrumentally mediated in a way parallel to Galileo's telescopic capture of mountains on their Moon. And, if this time-stop capacity of the technology can capture a horse's gait, the trajectory of even faster time-stop photography follows quickly. By 1888 time-stop photography had improved to the extent that the Mach brothers produced the first evidence of shock waves by photographing a speeding bullet. In this case, the photo showed that the bullet itself penetrated its target, not "compressed air," which was until then believed to advance before the projectile and cause injury (42–43).

Here we have illustrations of an early *perceptual hermeneutic* process which yields visually clear, repeatable, convincing Gestalts of the phenomena described. At this level, however, there is a "realism" of visual result which retains, albeit in a time-altered form, a kind of visual *isomorphism* which is a variant upon ordinary perception. It is thus less "textlike" than many other variants which develop later.

The visual isomorphism of early still photography was also limited to surface phenomena, although with a sense of frozen "realism" which shocked the artists and even transformed their own practices.[19] The physiognomy of faces and things was precise and detailed. The stoppage of time produced a *repeatable image of a thing*, which could be analytically observed and returned to time and again.

A second trajectory, however, was opened by the invention of the X-ray process in 1896. Here the "insides" of things could be depicted. Surfaces became transparent or disappeared altogether, and what had been "invisible" or, better, occluded became open to vision. X-ray photos were not so novel as to be the first interior depictions; we have already noted the invention of the "exploded diagram" style practiced by da Vinci and Vesalius. And one could also note that various indigenous art, such as that of Arnhemland Aborigines and Inuit, had an "X-ray" style of drawing which sometimes showed the interiors of animals. But the X-ray photo did to its objects what still photography had done to surfaces—it introduced a time-stop, "realistic" depiction of interior features. In this case, however, the X-ray image not only depicts differently, but produces its images as a "shadow." The X-rays pass through the object, with some stopped by or reduced by resistant material—in early body X-rays, primarily bones.[20]

Moving rapidly, once again a trajectory may be noted, one which followed ever more distinct depiction in the development of the imaging technologies: today's MRI scan, CT tomography, PET scans, and sonograms all are variants upon the depiction of interiorities. Each of these processes not only does its depicting by different means but also produces different visual selectivities which vary what is more or less transparent and what is more or less opaque. (I shall return to these processes in more detail.) This continues to illustrate the inscription or visualization process which constitutes the perceptual hermeneutic style of science.

A third trajectory in visualization is one which continues from the earliest days of optical instrumentation: the movement to the ever more microscopic (and macroscopic) entities. The microscope was much later to find its usefulness within science than the telescope. As Ian Hacking has pointed out, as late as 1800 Xavier Bichet refused to allow a microscope in his lab, arguing that "When people observe in conditions of obscurity each sees in his own way and according as he is affected."[21] In part, this had to do with the features of the things to be observed. Many micro-organisms were translucent or transparent and hard to make stand out even as figures against the often fluid grounds within which they moved. When another device which "prepared" the object for science was invented, *staining processes through aniline dyes,* the microscope could be more scientifically employed.[22]

The trajectory into the microscopic, of course, explodes in the nineteenth and twentieth centuries, with electron microscopes, scanning, tunneling processes, and on to the processes which even produce images of atoms and atom surface structure. Let me include here, too, the famous radio chrystallography which brought us DNA structure, as well as today's chromosome and genetic fingerprinting processes.

The counterpart, macro-imaging, occurs with astronomy and the "earth sciences" which develop the measuring processes noted in chapter 4, concerning "whole Earth measurements." While each trajectory follows a different, exploitable image strategy, the result retains the Gestalt-charactered visualization which is a favored perceptual object within science.

The examples noted above all retain repeatable Gestalt, visualizable, and in various degrees, isomorphic, features. This is a specialized mode of perception and perceptual hermeneutics which plays an important role within science, but which also locates this set of practices within a now complicated lifeworld.

B. "Textlike" visualizations

I shall now turn to a related, but different, set of visualizations, visualizations which bear much stronger relations to what can be taken as "textlike" features. Again, Latour is relevant: if the laboratory is science's *scriptorium,* the place where inscriptions are produced, then some of the production is distinctly textlike. A standard text, of course, is perceived. But to understand it one must call upon a specific hermeneutic practice—*reading,* and the skills which go into reading.

Once again, it would be tempting to follow out in more detail some of the history of the writings which have made up our civilizational histories (and which characterize the postmodern penchant for textuality, following Derrida's *On Grammatology*). But as far as written texts are concerned, I want to note in passing only that the histories of writing have tended to converge into an ever narrowing set of choices: alphabetical, ideographic, and, for special purposes, simple pictographic forms. Related to this shrinkage of historical forms, I also want simply to note in passing that science follows and exacerbates this trend within its own institutional form, so much so that its dominantly alphabetic actual text preference is even more clearly narrowed to the emerging dominance of English as "the" scientific natural language.

And there is plenty of "text" in this sense within science. The proliferation of journals, electronic publications, books, and the range of texts produced is obvious enough. These texts, however, always remain secondary or tertiary with respect to science, as we have seen from Latour. So this is not the textlike phenomenon I have in mind; instead, I am pointing to those analogues of texts which permeate science: charts, graphs, models, and the whole range of "readable" inscriptions which remain visual, but which are no longer isomorphic with the referent objects or "things themselves."

Were we to arrange the textlike inscriptions along a continuum, from the closest analogue to the farthest and the most abstractly disanalogous, one would find some vague replication of the history of writing. Historians of alphabetic writing, for example, have often traced the letters of alphabetic writing to earlier *pictographic* items in pre-alphabetic inscriptions. Oscar Ogg, for example, shows that our current letter "A" derives from an inverted pictograph of a bull image.[23] Earlier hieroglyphic inscriptions could serve double purposes: as an analog image of the depicted animal or as the representation of a particular phoneme in the alphabetic sense.

The vestigial analog quality noted in the history of writing also occurs in scientific graphics: for example, a typical "translation" technology occurs in oscillography. If a voice is being patterned on an oscilloscope, the sound is "translated" into a moving, squiggly line on the scope. Each sound produces a recognizable squiggle, which highly skilled technicians can often actually "read."[24] The squiggle is no more, nor no less, "like" the sound made than the letter is within a text, but the technical "hermeneutic" can read back to the referent. As the abstraction progresses, often purposefully so that a higher degree of *graphic Gestalt* can be visualized, the reading-perception becomes highly efficient. "Spikes" on a graph, anomalies, upward or downward scatters—all have immediate significance to the "reader" of this scientific "text." Here is a hermeneutic process within normal science. And it remains visualizable and carries now in a more textlike context the repeatable Gestalt qualities noted above as part of the lingua franca of this style of hermeneutic.

Older instrumentation often was straightforward analogous to the phenomenon being measured. For example, columns of mercury within a thermometer embodied the "higher" and "lower" temperatures shown. Or, if a container was enclosed, a glass tube on the outside with piping to the inside could show the amount of liquid therein. Even moves to digital or numeric dials often followed analog representations.

Finally, although I am not attempting comprehensiveness in this location of hermeneutics in my "weak program" within scientific practice, I want to conclude with some *conventions* which also serve to enhance the textlike reading perceptions. Graphs come with conventions: up and down for high and low temperatures or intensities; with the range of the growing uses of "false color" imagery, rainbow spectrum conventions are followed again for intensities, and so on. All of this functions "like" a reading process, a visual hermeneutics which retains its visualizations, but which takes textlike directions.

3. Summary

In the weak program I have been following to this point, I have chosen science activities which clearly display their hermeneutic features. These, I have asserted, include a preference for visualization as the chosen sensory mode for getting to the things. But, rather than serving simply as a reduction of perceptual richness by way of a monosensory abstraction, visualization has been developed in a hermeneutic fashion—akin to "writing" insofar as writing is also a visual display. Thus, if science is separate from the lifeworld, it is so in precisely the same way that writing would not be included as a lifeworld factor.

Second, I have held that the process within science practice which prepares things for visualization includes the instrumentarium, the array of technologies which can produce the display, depiction, graphing, or other visualizable result which brings the scientific object "into view." (I am not arguing, as some have, that *only* instrumentally prepared objects may be considered to be scientific objects. But, in the complex late Modern sciences, instrumentation is virtually omnipresent and dominant when compared to the older sciences and their observational practices.)

Third, I am not arguing that these clearly hermeneutic practices within science *exhaust* the notion of science. I have not dealt with the role of mathematization, with forms of intervention which do not always yield visualizable results, or the need to take apart the objects of science, to analyze things. And I do not mean to imply that these factors are also important to science. Rather, I have been making, so far, the weak case that there are important hermeneutic dimensions to science, especially relevant to the final production of *scientific knowledge.* In short, hermeneutics occurs inside, within, science itself.

13

Technoconstruction

Moving now from the implicit hermeneutics within science praxis to the more complex practices—increasingly technologically embodied and instrumentally constructed—we are ready to take note of a stronger program.

1. The "Strong Program": Hermeneutic Sophistication

In the "weak program" I chose to outline what could easily be recognized as hermeneutic features operative within science. As I now turn to a stronger program, I shall continue to examine certain extant features within science practice which relate to hermeneutic activity, but I shall increasingly turn here to forefront modes of investigation which drive the sciences closer to a postmodern variant upon hermeneutics.

A. Whole body perception

It is, however, also time to introduce more fully, albeit sketchily, a phenomenological understanding of perception in action. This approach will be recognizably close to the theory of perception developed by Merleau-Ponty, although taken in directions which include stronger aspects of multistability and polymorphy, which earlier investigations of my own developed.

1. I have already noted some perceptual *Gestalt features*, including the presentation of a perceptual field, within which figure/ground phenomena may be elicited. Following a largely Merleau-Pontean approach, one notes that fields are always complexly structured, open to a wide variety of intentional interests, and bounded by a horizonal limit. Science, I have claimed, in its particular style of knowledge construction, has developed a visualist hermeneutic which in the contemporary sense has

fulfilled its interests through *imagery* constituted instrumentally or technologically. The role of repeatable, Gestalt patterns, in both isomorphic and graphic directions, is the epistemological product of this part of the quest for knowledge.

2. In a strong sense, all sensory fields, whether focused upon in reduced "monosensory" fashion, or as ordinarily presented in synthesized and multidimensional fashion, are *perspectival* and concretely *spatial-temporal.* Reflexively, the embodied "here" of the observer not only may be noted but is a constant in all sensory perspectivalism. This constant may be enhanced only by producing a string of interrelated perspectives, or by shifting into multiperspectival modes of observation. The "ideal observer," a "god's eye view," and nonperspectivalism do not enter a phenomenology of perception.

3. However, while a body perspective relative to the perceptual field or "world" is a constant, both the field and the body are *polymorphic* and *multistable.* In my work in this area, I have shown that multistability is a feature of virtually every perceptual configuration (and the same applies to the extensions and transformations of perception through instrumentation), and that the interrelation of bodily (microperceptual features) and cultural significations (macroperceptual features) makes the polymorphy even more complex. *There is no perception without embodiment; but all embodiment is culturally and praxically situated and saturated.*

4. While I have sometimes emphasized spatial transformations (Galileo's telescope) in contrast to temporal transformations (still photography), all perceptual spatiality is *spatial-temporal.* This space-time configuration may be shown with different effects, as in contrasts between visual repeatability and auditory patterning, but is a constant of all perceptions.

5. All perceptual phenomena are synesthetic and multidimensioned. The "monosensory" is an abstraction—although useful and possible to forefront—and simply does not occur in the experience of the "lived body" (*corps vécu*). The same applies, although not always noted, in our science examples. I will say more below on this feature of perception. The issue of the "monosensory" is particularly acute with respect to the technological embodiments of science, since instruments (not bodies) may be "monosensory." Again, we reach a contemporary impasse which has been overcome only in part. Either we turn ingenious in the ways of "translating" the spectrum of perceptual phenomena into a visual hermeneutic—perhaps the dominant current form of knowledge construction in science—or we find ways of enhancing our instrumental reductions through variant instruments or new modes of perceptual transformation (I am pointing to "virtual reality" developments here).

A "strong program," I am hinting, may entail the need for breakthroughs whereby a fuller sense of human embodiment may be brought into play in scientific investigations. Whereas the current, largely visualist hermeneutic within science may be the most sophisticated such mode of knowledge construction to date, it remains short of its full potential were "whole body" knowledge made equally possible. This would be a second step toward the incorporation of lifeworld structures within science praxis.

B. Instrumental phenomenological variations

In the voice metaphor I used to describe the investigation of things, I noted that the "giving of a voice" entails, actually, the production of a "duet" at the least. But this also means that different soundings may be produced, either in sequence or in array, by the applications of different instruments. This is a material process which incorporates the practice of "phenomenological variations" along with the intervention within which a thing is given a voice. This practice is an increasing part of science practice and is apparent in the emergence of a suite of new disciplines which today produce an ever more rapid set of revolutions in understanding or of more frequent "paradigm shifts."

I use this terminology because it is a theme which regularly occurs in science reporting. A Kuhnian frame is often cast over the virtually weekly breakthroughs which are reported in *Science, Scientific American, Nature,* and other magazines. Challenges to the "standard view" are common. I shall look at a small sample of these while relating the challenges to the instrumental embodiments which bring about the "facts" of the challenges. Here the focus is upon *multiple instrumental arrays* which have different parameters in current science investigation.

1. *Multiple new instruments/more new things.* The development of multivariant instruments has often led to increased peopling of the discipline's objects. And much of this explosion of scientific ontology has occurred since the mid–twentieth century. This is so dramatically the case that one could draw a timeline just after World War II, around 1950 for most instrumentation, and determine new forms for many science disciplines. For example, in astronomy, until this century, the dominant investigative instrumentation was limited to *optical* technologies and thus restricted to the things which produce *light.* With the development of *radio-telescopy,* based upon technologies developed in World War II—as so many fields besides astronomy also experienced—the field expanded to the forms of microradiation which occur along spectra beyond the

bounds of visible light. *The New Astronomy* makes this "revolution" obvious. The editors note that

> The range of light is surprisingly limited. It includes only radiation with wavelengths 30 per cent shorter to 30 per cent longer than the wavelength to which our eyes are most sensitive. The new astronomy covers radiation from extremes which have wavelengths less and one thousand-millionth as long, in the case of the shortest gamma rays, to over a hundred million times longer for the longest radio waves. To make an analogy with sound, traditional astronomy was an effort to understand the symphony of the Universe with ears which could hear only middle C and the two notes immediately adjacent.[1]

Without noting which instruments came first, second, and so on, expanding out from visible light, first to ultraviolet on one side, and infrared on the other, now reaching into the previously invisible-to-eyeball perceptions, but still within the spectrum of optical light waves, the first expansion into invisible light range occurs through types of "translation" technologies as I have called them. The usual tactic here is to "constitute" into a *visible depiction* the invisible light by using some convention of *false color* depiction.

The same tactic, of course, is used once the light spectrum itself is exceeded. While some discoveries in radio astronomy were made by *listening* to the radio "hiss" of background radiation, it was not long before the gamma-to-radio wavelengths beyond optical capacities were also "translated" into visible displays.

With this new instrumentation, the heavens begin to show phenomena previously unknown but which are familiar today: highly active magnetic gas clouds, radio sources still invisible, star births, supernovas, newly discovered superplanets, evidence of black holes, and the like. The new astronomy takes us closer and closer to the "birth" of the universe. More instruments produce more phenomena, more "things" within the universe.

The same trajectory can be found in many other science disciplines, but for brevity's sake I shall leave this particular example as sufficient here.

2. *Many instruments/the same thing.* Another variant, now virtually standard in usage, is to apply a range of instruments which measure different processes by different means to the same object. Medical imaging is a good example here. If some feature of the brain is to be investigated, perhaps to try to determine without surgical intrusion whether a

formation is malignant or not, multiple instrumentation is now available to enhance the interpretation of the phenonomenon. A recent history of medical imaging, *Naked to the Bone,* traces the imaging technologies from the inception of X-rays (1896) to the present.

As noted above, X-rays allowed the first technologized making of the body into a "transparent" object. It followed the pattern noted of preparing the phenomenon for a scientific "reading" or perception. Early development entailed—to today's retrospective horror—long exposures, sometimes over an hour, to get barely "readable" images sometimes called "shadowgraphs." This is because X-ray imaging relies upon radiation sent through the object to a plate, and thus the degrees of material resistance cast "shadows" which form the "picture." The earliest problems focused upon getting clearer and clearer images.[2]

I have noted that the microscope became useful only when the specimen could be prepared for "reading" through a dye process which enhanced contrasting or differentiated structures (in the micro-organism). This image enhancement began to occur in conjunction with X-rays as early as 1911 with the use of *radioactive tracers* which were ingested or injected into the patient. This was the beginning of nuclear medicine (304).[3]

Paralleling X-ray technologies, *ultrasound* began to be explored with the first brain images produced in 1937 (310). The quality of this imaging, however, remained poor since bone tended to reduce what could be "seen" through this sounding probe. (As with all technologies, it takes some time before the range of usefulness is discovered appropriate to the medium. In the case of ultrasound, soft tissue is a better and easier-to-define target.)

But, even later than the new astronomy, the new medical imaging does not actually proliferate into its present mode until the 1970s. Then, in 1971 and 1972, several patents and patent attempts are made for magnetic resonance processes (MRIs) (314). These processes produce imagery by measuring molecular resonances within the body itself. At the same moment, the first use of the computer—*as a hermeneutic instrument*—comes into play with the refinement of *computer-assisted tomography* (CT scanning). Here highly focused X-ray beams are sent through the object (brains, at first), and the data are stored and reconstructed through computer calculations and processes. Computer-"constructed" imaging, of course, began in the space program with the need to turn data into depictions. Kevles notes that

> After the Apollo missions sent back computer-reconstructed pictures of the moon, it did not stretch the imagination to propose that computers

could reconstruct the images of the interior of the body, which, like pictures from space, could be manipulated in terms of color and displayed on a personal video moniter. (143–44)

Here mathematics and imagery or constructed perceivable depictions meet. I claim this is important to a strong hermeneutic program in understanding science.

By 1975 the practical use of positron emissions is captured in the PET scan process. These emissions (from positrons within the object) are made visible (314). This imagery has never attained the detail and clarity of the above technologies but has some advantage in a dynamic situation when compared to the "stills" which are produced by all but ultrasound processes (and which also are limited in clarity). Thus living brain *functions* can be seen through PET instrumentation. Then, in the 1990s, functional MRI and more sophisticated computer tomographic processes place us into the rotatable, three-dimensional depictions which can be "built up" or "deconstructed" at command, and the era of the *whole body image* is attained (314).

While each of these processes can show different phenomena, the multiple use is such that ever more complete analysis can also be made of single objects, such as tumors, which can be "seen" with differences indicating malignancy or benignity. This, again, illustrates the ever more complex ways in which science instrumentation produces a visible result, a visual hermeneutics which is the "script" of its interpretive activity.

3. *Many instruments/convergent confirmations.* Another variant upon the multiple instrument technique is to use a multiplicity of processes to check—for example, dating—for greater agreement. In a recent dating of Java *homo erectus* skulls, uranium series dating of teeth, carbon 14, and electron spin resonance techniques were all used to establish dates much more recent (27,000 B.P. +/- 53,000 B.P. for different skulls) than previously determined and thus found *homo erectus* to be co-extant with *homo sapiens sapiens.*[4] And, with the recent discovery of 400,000 B.P. javelin-like spears in Germany, one adds thermo-luminescence techniques to establish this new date for human habitation in Europe, at least double or triple the previously suspected earliest date for humans there. (Similar finds, now dated 350,000 B.P. in Siberia, and 300,000+ B.P. finds in Spain, all within 1996–97 discovery parameters, evidence this antiquity.)[5]

4. *Single instrument (or instrumental technique)/widespread multiple results.* Here perhaps one of the most widely used new techniques involves DNA "fingerprinting," which is now used in everything from forensics (rapists and murderers both convicted and found innocent and released), to pushing dates back for human migrations or origins. (The

"reading" process which goes with DNA identification entails matching pairs and includes visualizations once again.) *Scientific American* has recently reported that DNA tracing now shows that human migrations to the Americas may go back to 34,000 years (not far from the dates claimed for one South American site, claimed to be 38,000 B.P., which with respect to physical data remain doubtful), with other waves at 15,000 B.P., to more recent waves, included a set of Pacific originated populace around 6,500 B.P.)[6] The now widely cited DNA claims for the origins of *homo sapiens* between 200,000 and 100,000 B.P. is virtually a commonplace.

DNA fingerprinting has also been used in the various biological sciences to establish parentage compared to behavioral mating practices. One result is to have discovered that many previously believed-to-be "monogamous" species are, in fact, not. Similarly, the "Alpha Male" presumed successful at conveying his genes within territorial species has been shown to be less dominantly the case than previously believed.[7]

5. *Multiple instruments/new disciplines.* Beginning with DNA (mitochondrial DNA matching) again, the application of this technique has given rise to what is today called "ancient DNA" studies, with one recent result developed in Germany this year, which purports to show that Neanderthals could not have interbred with modern humans, due to the different genetic makeup of these hominids who coexisted (for a time) with modern humans.[8]

Then, returning to variants upon imaging, the new resources for such disciplines as archeology produce much more thorough "picturing" of ancient sites, activities, and relations to changes in environmental factors. Again, drawing from *Scientific American*—one can draw similar examples from virtually every issue of this and similar science-oriented magazines—the array of instruments now available produces, literally, an "in-depth" depiction of the human past. In part, now drawing from uses originally developed for military purposes, imaging from (1) *Landsat,* which used digital imaging and multispectral scanners from the 1970s, to (2) refinements for *Landsats 4* and *5* in the 1980s, which extended and refined the imaging and expanded to infrared and thermal scanning, (3) to *SPOT,* which added linear-array technologies to further refine imaging, to (4) imaging radar, which actually penetrates below surfaces to reveal details, to (5) *Corona,* which provides spectrographic imagery (recently declassified), the modern archeologist, particularly desert archeologists, can get full-array depictions of lost cities, ancient roads, walls, and the like from remote sensing used now.[9]

Once located, instrumentation on Earth includes (1) electromagnetic sounding equipment, which penetrates up to six meters into the earth, (2) ground-penetrating radar, which goes down to ten meters,

(3) magnetometers, which can detect such artifacts as hearths, (4) resistivity instruments, which detect different densities and thus may be used to locate artifacts, and (5) seismic instruments, which can penetrate deeper than any of the above instruments. At a recent meeting in Mexico, an anthropologist reported to me that a magnetometer survey of northern Mexico has shown there to be possibly as many as 86,000 buried pyramids (similar to the largest, Cholula, although the remainder are smaller).[10]

In short, the proliferation of instrumentation, particularly that which yields imagery, is radical and contemporary and can now yield degrees and spans of three-dimensional imagery which includes all three of the image breakthroughs previously noted: early optics magnified the micro- and macro-aspects of barely noted or totally unnoted phenomena through magnification (telescopy and microscopy) but remained bound to the limits of the optically visible, by producing "up close" previously distant phenomena.

Photography increased the detail and isomorphy of imaging in a repeatable produced image, which could then be studied more intensely since "fixed" for observation. It could also be manipulated by "blowups" and other techniques, to show features which needed enhancement. Then, with X-rays, followed by other interior-producing imagery, the possibilities were outlined for the contemporary arrays by which image surfaces are made transparent so that one may see interiors. Then add the instruments which expand thoroughly beyond the previously visible and which now go into previously invisible phenomena through the various spectra which are "translated" into visible images for human observation.

Here we have the decisive difference between ancient science and Modern science: Democritus claimed that phenomena, such as atoms, not only were in fact imperceptible, but were *in principle* inperceptible. A modern, technologized science returns to Democritus to the "in fact" only—that is, the atomic is invisible only until we can come up with the technology which can make it visible. This, I claim, is an instrumental *visual hermeneutics.*

C. Technoconstitution

In the reconstrual of science which I have been following here, I have argued that late Modern science has developed a complex and sophisticated system of *visual hermeneutics.* Within that visualist system, its "proofs" are focused around the things seen. But, also, things are never just or merely seen—the things are *prepared* or made "readable." Scientifically, things are (typically, but not exclusively) *instrumentally mediated,* and the "proof" is often a *depiction* or image.

Interestingly, if in ordinary experience there is a level of *naive realism* where things are taken simply to be what they are seen to be, similarly, within imaging there is at least the temptation to an imaging naive realism. That naïveté revolves around the intuitive taking of the image to be "like" or to "represent" an original (which would be seen in unmediated and eyeball perceptions). In short, "truth" is taken to be some kind of *isomorphism* between the depiction and the object.

By putting the issue in precisely that way, there are many traps which are set which could lead us back into the issues of modern (that is, Cartesian or seventeenth- and eighteenth-century) epistemologies. But to tackle these would lead us into a detour of some length. It would entail deconstructing "copy theory" from Plato on, deconstructing "representationalism" as the modern version of copy theory, before finally arriving at a more "postmodern" theory which entails both a theory of relativistic intentionality, a notion of perspectivalism, and an understanding of instrumental mediations as they operate within a phenomenological context. (I have addressed these issues in some degree in essays which preceded my formulation here.).[11] I simply want to avoid these traps.

To do so, I shall continue to interpret science in terms of a visual hermeneutics, embodied within an instrumentally realistic—but *critical*—framework in which instruments mediate perceptions. The device I shall now develop will fall within an idealized "history" of imaging, which, while containing actual chronologically recognizable features, emphasizes patterns of *learning to see.*

1. Isomorphic visions

The first pattern is one which falls into one type of *initial isomorphism* within imaging. As a technical problem, it is the problem of getting to a "clear and distinct" image. Imaging technologies do not just happen, they develop. And in the development there is a dialectic between the instrument and the user in which both a learning-to-see meets an elimination-of-bugs in the technical development. This pattern is one which, in most abstract and general terms, moves from initial "fuzziness" and ambiguity to greater degrees of clarity and distinctness.

Histories of the telescope, the microscope, photography, and X-rays (and, by extension, all the other imaging processes as well) are well documented with respect to this learning-to-see. Galileo, our quasi-mythical founder of early Modern science, was well aware of the need to teach telescopic vision, and of the problems which existed—although he eventually proclaimed the *superiority* of instrumentally mediated vision over ordinary vision. The church fathers, however, did have a point about how to take what was seen through the telescope. Not all of Galileo's observations were clear and easily seen by "any man." The same problem reemerged in the nineteenth century through the observations of Giovanni Schiaparelli, who gave the term the "canals of Mars." Schiaparelli was a well-known astronomer who had made a number of important discoveries, particularly with respect to asteroids and meteor swarms (because, in part, he had a much better telescope than Galileo). But in noting "canali"—which should have been translated into "channels"—which were taken to be "canals"—he helped stimulate the speculations about life on Mars. But neither channels nor canals existed—these, too, were instrumental artifacts.[12]

The dialectic between learning and technical refinement, in the successful cases, eventually leads to the production of clear and distinct images and to quick and easy learning. These twin attainments, however, cover over and often occlude the history and struggle which preceded the final plateaus of relative perfections. Thus, as in the previous illustrations concerning my guests and our Vermont observations of the Moon, once focused and set, it literally takes only instants before one can recognize nameable features of its surface. The "aha phenomenon," in short, is virtually immediate today because it is made possible by the advanced technologies. That instantaneity is an accreted result of the hidden history of learning-to-see and its accompanying technical debugging process.

This same pattern occurred with the microscope. Although microorganisms never before seen were detected early, the continued problems of attaining clear and distinct microscopic vision was so difficult that it did not allow the microscope to be accepted into ordinary scientific practice until the nineteenth century. Again, the dialectic of learning how and what to see meets the gradual technical improvement concerning lenses and focusing devices, and finally the application of dying procedures to the things themselves. (This is an overt example of preparing a thing to become a scientific or "readable" object!)

Photography stands in interesting contrast to microscopy—if it took a couple of centuries for microscopy to become accepted for scientifically acceptable depiction, photography was much faster to win the same position. From Niepce's first "fixed" image in 1826, to the more widely accepted date of 1839 for Daguerre's first images, it was less than a half century until, as Bettyann Kevles notes in her history of medical imaging, "By the 1890's photographs had become the standard recorders of objective scientific truth."[13]

The same pattern occurs, but with even greater speed, in the history of early X-rays. In publicizing his new invention, Wilhelm Röntgen made copies of the X-ray of his wife's hand, which showed the bones of her fingers and the large ring which she wore, and sent these to his colleagues across Europe as evidence of his new process. That X-ray (with a long exposure time) was fuzzy, and while easily recognizable as a skeletal hand and ring, contrasts starkly with the radiograph made by Michael Pupin of Prescott Butler's shot-filled (shotgun-injured) hand later the same year (21, 35). X-rays, duplicated across Europe and America almost immediately after Röntgen's invention, were used "scientifically" from the beginning.

The acceleration of acceptance time (of the learning-technical vision dialectic) similarly applies to the recent histories of imaging, which include, as above, sonograms (1937) and MRIs (1971) in medicine, of remote imaging since the *Tiros* satellite (1965), or of digitally transmitted and reconstructed images from *Mariner 4* (1965) in Earth and space science.

All of the above samples, however, remain within the range of the possible "naive image realism" of visual isomorphisms in which the objects are easily recognizable, even when new to the observer's vision. (Even if Röntgen had never before seen a "transparent hand" as in the case of his wife's ringed fingers, it was "obvious" from the first glimpse what was seen.) The pattern of making clear is an obvious trajectory. Yet we are not quite ready to leave the realm of the isomorphic.

How does one make "clearer" what is initially "fuzzy"? The answer lies in forms of manipulation, what I shall call *image reconstruction.* The techniques are multiple: enlargements (through trajectories of magnification noted before), enhancements (where one focuses in upon particular features and finds ways to make these stand out), contrasts (by heightening or lessening features of or around the objects), and so on. In my examinations I shall try not to be comprehensive, but to remain within the ranges of familiarity (to at least the educated amateur) concerning contemporary imaging. All of these manipulations can and do occur within and associated with simply isomorphic imaging and, for that matter, within its earlier range of black-and-white coloring. Histories of the technical developments which go with each of these techniques are available today and provide fascinating background to the rise of scientific visualism.

The moral of the story is images don't just occur. They are made. But, once made—assuming the requisite clarity and accuracy and certification of origin, etc.—they may then be taken as "proofs" within the visual hermeneutics of a scientific "visual reading." We are, in a sense, still within a Latourean laboratory.

2. Translation techniques

Much of what can follow in this next step has already been suggested within the realm of the isomorphic. But what I want to point to here is the use in late Modern science of visual techniques which begin ever more radically to vary *away from the isomorphic.*

One of these variables is—if it could be called that—simply the variable use of *color.* Returning to early optics, whatever Galileo or Leeuwenhoek saw, they saw in "true color." And, as we have seen, sometimes that itself was a problem. The transparent and translucent micro-organisms in "true color" were difficult to see. With aniline dyes, we have an early use of "false color." To make the thing into a scientific or "readable" object, we intervene and create a "horse of a different color." "False coloring" becomes a standard technique within scientific visual hermeneutics.

The move away from isomorphism, taken here in gradual steps which do not necessarily match chronologically what happened in the history of science, may also move away from the limits of ordinary perception. As noted above, the "new" late Modern astronomy of midcentury to the present was suddenly infused with a much wider stretch of celestial "reality" once it moved beyond optical and visible limits into, first, the humanly invisible ranges of the still optical or light itself, in the ranges of the infrared and ultraviolet. The instrumentation developed was what I have been calling a *translation technology* in that the patterns which are recordable on the instrumentation can be rendered by "false coloring" into visible images. This same technique was extended later to the full wave spectrum now available from gamma rays (short waves) through the optical to radio waves (long waves), which are rendered in the standard visually gestaltable, but false color, depictions in astronomy. All this is part of the highly technologized, instrumentalized visual hermeneutics which makes the larger range of celestial things into seeable scientific objects.

The "realism" here—and I hold that it is a realism—is a Hacking-style realism: if the things are "paintable"[14] (or "imagable") with respect to what the instruments detect as effects which will not go away, then they are "real." But they have been *made visible* precisely through the technological constructions which mediate them.

3. Higher level construction

Within the limits of the strong program, I now want to take only two more steps: I am purposely going to limit this attempt to reconstrue science praxis as hermeneutic to contemporary imaging processes which make (natural) things into scientific, and thus "readable," visual objects. I am not going to address the related, but secondary, visual process which entails *modeling.* That process which utilizes the computer as a hermeneutic device is clearly of philosophical interest, but I shall stop short of entering that territory here.

Computers, of course, are integral to many of the imaging processes we have already mentioned. Medical *tomography* (MRI, PET, fMRI, etc.) entails computer capacities to store and construct images. What is a visual Gestalt is built up from linear processes which produce data which have to be "constructed" by the computer. Similarly, the digitally transmitted imagery from distance sensing in satellite, space, and other remote imaging processes also has necessary computer uses. Much of contemporary imaging is computer embodied. And computers open the ways to much more flexible, complex, and manipulable imaging than any previous technology. For the purposes here, however, they will remain simply part of the "black boxes" which produce images which mediate perceptions.

The two higher-level constructive activities I want to point to here entail, first, the refinement of imaging which can be attained through specifically recognizing our technologies as mediating technologies which, in turn, must take into account the "medium" through which they are imaging. I turn again to astronomical imaging: the *Hubble* space telescope has recently captured the most public attention, but it is but one of the instrumental variations which are today exploring the celestial realms.

The advantage *Hubble* has is that it is positioned beyond the effects of the atmosphere with its distortions and interferences—the clarity of *Hubble* vision in this sense is due in part to its extra-Earthly perspective. (Science buffs will recall that at launch it had several defects in operation which were subsequently fixed—thus placing the *Hubble* in the usual pattern of needing technical adjustment to make its images clear!) But, in part by now being able to (phenomenologically) vary *Hubble* with Earth-bound optical telescopes, the move to enhancing Earth-bound telescopy through computer compensations has become possible. Astronomy is moving toward technoconstructions which can account for atmospheric distortions "on the spot" through a combination of laser targeting and computer enhancements. Earth-bound telescopes are today being given new life through these hi-tech upgrades which "read" atmospheric distortions and "erase" these processes which can make clearer new "readable" images. *Science* regularly publishes an "imaging" issue devoted to updating what is taken as the state-of-the-art in imaging (in 1997 it was the 27 June issue). A description of how one "undoes the atmosphere" is included, which entails computer reconstructions, telescopes in tandem, and adaptive optics. This process, *Science* claims, "combat[s] the warping effects of gravity on their giant mirrors . . . reclaims images from the ravages of the atmosphere . . . [and] precisely undoes the atmospheric distortions."[15]

But alongside *Hubble* are the other variants: the infrared space observatory, the *Cosmic Background Explorer*, and other satellite instrumentation which produces imagery from the nonoptical sources. All these technologies are variants upon the same multidimensioned variables which produce readable images, or make things into readable scientific objects.

The final set of instrumental productions I wish to note are the *composites* which produce variants upon "wholes." Chapter 4 deals with "whole Earth measurements" which constitute one realm of composite imagery. To determine whether or not sea levels are rising overall, the composite imagery produced combines (1) multiple satellite photo imagery, (2) Earth-bound measurements (such as buoys, laser measurements, and land markers), and (3) computer averaging processes to produce a depiction (false colored) which can, in comparing time slices, show how much the oceans have risen. The composite depiction displays a flat-projection map of the Earth with level plateaus in false color spectra which can be compared between years, decades, and so on.

Similar processes occur in medical imaging. The "whole body imagery" available today on the internet is the result of two full-body "image autopsies," one each of a male and a female, whose bodies through tomographic processes may be seen in whatever "slice" one wishes. The linear processes of tomography show, slice-by-slice, vertically, horizontally, or in larger scans, the full bodies of the corpses used. The dimensions can be rotated, realigned, sectioned, and so on. Tomography also allows one to "peel," layer by layer, the object imaged—from skin, to networked blood vessels, to bones, and so on. (Both the whole Earth and whole body images are probably among the world's most expensive "pictures.") Moreover, all the manipulations which entail enhancements, contrasts, colorings, translations, and the like are utilized in these "virtual" images. Yet, while these virtual "realities" are different from the examination of any actual cadaver, they clearly belong to the visual hermeneutics of science in the strong sense. Things have been prepared to be seen, to be "read" within the complex set of instrumentally delivered visibilities of scientific imaging.

14

Beyond Visualism

The program I have followed here is a focused one: I am attempting to reframe an understanding of science in its *hermeneutic dimensions.* I am not claiming that science is exclusively hermeneutic (that would be a reductionism which could be a counterpart to the positivist reduction of science to a theory-generating machine) or that it lacks other dimensions. What I am claiming is that by reframing our understanding of science in terms of interpreting much of its praxis as hermeneutic, we gain certain insights into those operations.

I have argued that science includes within its praxis many features which are hermeneutic in kind. But I have also argued that science's version of a hermeneutics is of a particular and peculiar kind. It is not primarily linguistic or propositional, although there are versions of "reading" contained within its practice. It is, rather, perceptually oriented—but more, primarily *visually* oriented in its hermeneutic dimension. It has developed a complex, sophisticated *visual hermeneutics* which retains the advantages of visual perception but has infused these with a world of things never before included in naked, bodily perception.

1. Within Vision

The visualizations of a scientific hermeneutics take several directions and vectors: often its visual displays retain some degree of isomorphism with ordinary visual perceptions. But it also succeeds in varying away from these while retaining the Gestalt qualities of depictability. In a second vector, it also develops visualizations which are more "textlike" and no longer analogous to perceived objects. These textlike or *graphic* visualizations are "read through" as texts are read through.

Although a visual scientific hermeneutic may include ways of seeing which entail ordinary bodily perceptions, direct and directed upon things (particularly in such macrosciences as ecology, various fields of biology, field studies of many kinds, etc.), increasingly the visualizations are embodied in *technologies* or instruments which prepare and produce the visual product. Even field studies are often transcribed into charts, diagrams, and tables—along with photos of things observed. I have termed the technologically transformed products images since they are results of what today are called imaging technologies. Indeed, it is virtually exlusively through instruments that the human scientist can "see" beyond his ordinary eyeball perceptions "into" the micro- and macroworlds we now inhabit. I am thus interpreting technologies, the instrumentarium, as perception-mediating and perception-transforming devices.

I have quite deliberately limited my inquiry to this focus and, furthermore, sought to avoid the long discussions which are implied concerning the implicit *epistemology* which needs to be associated with this style of hermeneutics. (I will anticipate that it is *not* a modern epistemology of the style of the seventeenth and eighteenth centuries, which still lurks in the language of many of the sciences—and in some of the analytic philosophies. It is, rather, a more postmodern-plus-phenomenological epistemology which relies upon a body-relativity, perspective, world-constituting set of activities.)

Nor have I explored all the nuances, particularly the critical ones, which ultimately need to be included in this focus. For example, any good hermeneutic process will necessarily include critical means for sorting out falsehoods, dead ends, deceptions, and the like from within its "texts." (How long did the deliberate fiction of the "Donation of Constantine," for example, haunt European politics?) The same applies to the visualist hermeneutics of science. To reclaim one example set: the history of imaging is littered with "instrumental artifacts," that is, visual presentations which turn out to be effects of the instrument rather than of the referent thing. Galileo's famed "auras" around stars are one example (even if he took this as evidence of the superiority of instrumental vision and never discovered his error!). Schiaparelli's much later "canals" of Mars, so long part of the tradition of solar system lore, are another example. These phenomena no longer "exist"—they, like phlogiston, ether, and so many other "scientific objects," have ceased to be.

A second set of distortions revolves around the technologies of measurement with respect to *calibration*. If there is an error in calibration, the error will be projected upon the result. Again, the history of measurement is cluttered with such errors. Yet, with respect to both these problems, the contemporary use of multiple instruments and multivariant measurements helps to correct for both these problems. If one type of technology creates an "artifact," it is unlikely that the two or three or more other instruments which are aimed at the same phenomenon will duplicate the instrumental artifact. Similarly, a multivariant set of measurements can be regarded as more, rather than less, accurate when these different measurements converge (see examples above).

I have called this practice the use of "instrumental phenomenological variations" since the use of multiple variants varies the "profiles" of the phenomenon such that adequacy is attained only through sufficient variation. (Phenomenology is sometimes plagued with similar problems when "apodicticity" is claimed too quickly, or when variations have not been applied with radical enough imagination, which can lead to too quick closure upon the "essence" of a phenomenon.)

The juxtaposition of phenomenological critical techniques to scientific ones could continue quite indefinitely: for example, one phenomenological technique is to rely upon intersubjective checking since one investigator may well overlook or undervaluate something in the same phenomenon which another might "see." (This technique was honed under Herbert Spiegelberg's workshop approach to phenomenology at Washington University in the sixties.) The deliberate application of focus shifts and even figure/ground reversals is another technique which can produce critical results in both phenomenology and the sciences. Both of these techniques are variations upon a more broadly construed use of *multiperspectival* and *variational* applications (which are consonant with postmodern approaches as well).

The same critical extension of variational theory to instrumentally applied "perceptions" applies as well. Once one is beyond the limits of ordinary perception, while the task is to translate back into perceivability what is being discovered, many of the sensory aspects of this instrumental perceivability are cognate to different animal or biological "perceptual" processes. Military applications are instructive here: thermal imaging is an analogue to serpentine perception; enhanced-light technologies for nighttime vision are analogues to feline perception; wide-angle imaging and other "distortions" away from the forward-facing, core vision-focused human (and primate, canine, and other relatives with forward-enhanced vision in the animal kingdom) are often analogous to insect perceptions. The actual or empirical variations upon perception which, as such, extend beyond human horizons can be quasi-translated into the humanly perceivable. The "distortions" which remain are, in principle, no different from the imaginative perceptions artistic variations have often developed.

All of these comments and observations help reframe an understanding of contemporary science along a convergent trajectory between science praxis and the P-H traditions of epistemology. But they also remain bound to the perpetuation of a possible visualism in the hermeneutics employed within science. And this could be the determinate trajectory of scientific hermeneutics. Its limits have not been reached, and it has proven itself ever more ingenious in the production of the visual "proofs" it needs in its production of knowledge.

Why does it work so well? Part of the answer lies in the very analogue which lurks between science and hermeneutics. A visual hermeneutics in science is a material counterpart in a different context to the *invention of written language,* but not because science is a language analogue. Writing, more broadly construed as "inscription," is the *technologizing* of language. To write calls for (1) a translation of language into a nonspeech form—for simplicity's sake, the making of sounds *visible* (in phonetic writing), (2) combined with an instrumental mediation onto the machinery of the written (papyrus, paper, etc., with the use of stylus, pen, computer, etc.). Once these moves are made and perfected, the "explosion" of communication which occurred through the history of writing could happen. The materialization of language makes it repeatable and conveyable, even beyond its present or its place of origin. The *repetition* which is implied in being able to take up a text, over and over again, and with it the proliferation of interpretation, commentary, transport, translation, and the like lies at the birth of writing. But there is also a difference: texts, in one sense, change less than scientific objects (unless one takes deconstruction literally!).

Science's similar movement, into the "Latourean laboratory" whereby instruments become the scriptorum of things, by which the visual hermeneutics we have followed get developed and perfected, is its counterpart to written language. But the result is not primarily linguistic—it is rather the visualization of things as scientific objects. Thus, if science's version of a hermeneutic is visual, this visuality functions, not simply as a "reduction" as some have phrased the relationship between written and spoken language, but as the privileged dimension within which what is seen is understood. A "universal hermeneutics" is herein implied, but it is a hermeneutics of things, not of language alone.

2. Beyond Vision

Need science remain within a visual hermeneutic? In its broader praxis it is, of course, just as multidimensioned and bodily actional as any other human activity. Even in the expression of its interpretation, it has not always been limited to visualizations, in spite of its early vectors which enhanced that direction. As previously noted, auditory or acoustic perceptions, along with clearly acoustic instrumentation which produced auditory demonstrations (rather than visual ones), have periodically occurred. And, when acoustic instrumentation advances better or faster than visual instrumentation, then science follows its technological trajectories. Kevles notes that in the late nineteenth century—just before the success of the visual imaging processes—some individuals thought the future of knowledge production lay with acoustic technologies. The telephone (1876) and phonograph (1897) had sparked interest, and Edward Bellamy offers the suggestion that

> hearing is the sense of the future, and coming none too soon to rescue eyesight, which was "indeed terribly overburdened previous to the introduction of the phonograph." . . . In the early 1890's sound seemed to be the door to a technology explosion. In medicine the stethoscope had become a staple in the physician's black bag, and percussion of the chest had become a routine method of sounding out disease.[1]

These preceded the twentieth-century forms of auditory remote sensing (as in sonar and early radio-teloscopy). And, in biological field studies, the discovery and recording of subsurface sea life (whale songs, for example) are acoustically presented (but also displayed on voice oscillographs). Yet, these auditory isomorphisms eventually yielded to visualizations (oscilliscopes, visual radar and sonar, false color images of radio wave phenomena).

Similarly, there are rare instances in which blind scientists have excelled in fields through auditory and *tactile* perceptions rendered sharp in contrast to the inherent visualism of science. For example, the molluskologist Gerrat Zermeiz became an expert on evolution in mollusks through his tactile perception of virtually unseeable features of shells and their patterned changing configurations.[2]

Not even the metaphors are all visual. The emphasis upon intervention and manipulation, along with the "phallic" metaphors of probing, penetrating, and even wrestling with things, also belongs to the self-understanding of many scientists. In short, "whole body" action lies behind the full praxis of science just as it does for every other human endeavor. But, if the visual hermeneutic is the standarized expression of the understanding of things in science, could it be otherwise?

Here I enter speculative ground, and I shall do so by way of the state of instrumentation which is both available today, and which is under development. Perhaps the two most sophisticated sets of instrumentation which entail "sensing" belong to science—and the military. Since the end of the Cold War, one scientifically significant set of events lies in the transformation of previously classified and secret militarily produced knowledge and its adaptation by scientists today.

For example, one recently publicized set of releases has to do with the topography of ocean bottoms. Visual maps of detailed, three-dimensional projections now detail virtually the entirety of the oceans. The military need to know this for submarine warfare: Where could one hide? How could one find the enemy? The now-public results were produced by precisely the various multi-instrument and multivariant processes noted above. The "maps" now available are, again, *composites* of these processes.

The process begins with a wide-scan process via a *Geosat* satellite using a multibeam sonar. The ocean surface is imaged and then, done indirectly by calculating gravity effects (sea mounts cause gravitational attraction, etc.) which can be seen with a relief scale of 200 meters, followed by the use of vessel-towed, undersea multibeam sonar, the resolution improves to much greater precision (the rule of thumb is an inversion: wider scan, less resolution, but narrower scan takes more time and covers less territory). Finally, after using these "auditory" analogues, one turns to undersea photography via robot or submersible, for both confirmation and refinement. Once again, we have a composite result which is then reconstucted by computer processes and produces three-dimensional maps of undersea coastal, sea mount, or trench displays.[3]

In the case above, the result is, once again, a visual product: the three-dimensionally depicted bottom topography. This means all the various instruments, using the different processes, are "composed" into the final visual result—ingenious, but clearly within the systematic standard now dominant.

Could a multisensory "hermeneutic" be developed? This time I turn in this speculative venture to the instruments themselves. Technologies as mediating devices—media—are often more than monosensory. The optical and variant upon optical trajectory begun in early Modernity and made more complex and sophisticated in late Modernity is only one trajectory. A second trajectory is one which has sought to mimic the multidimensionality of synesthetic perception. There was a period in the late nineteenth century when acoustical techniques were often better than visual ones, particularly for interiors, but these were surpassed with the invention of the X-ray and subsequent transparancy-making techniques.

It was not science, however, which pushed for a multisensory approach: it was more from *entertainment*. Indeed, art and entertainment uses frequently *precede* ultimate scientific uses. I have already noted

Descartes's use of the camera obscura as model for both eye and ego. These devices, along with grids and projective devices from the Renaissance on, were used first by artists, but then later for accurate drawings of celestial phenomena or of botanical and other natural history items by scientists. All these devices produced hand-drawn images, later replaced largely by photographs. At first, photos were limited to stills. But by the turn of the century, photography had progressed beyond stills to "movies" (1895) albeit at first "silent" movies. Paralleling this development was the reproduction of sound in early phonographs ([1877], to which one must add the telephone [1876], early radio [1920s], and other auditory technologies). The breakthrough toward multisensory presentations came with the combination of sight and sound—the soundtracks now appended to and synchronized with the film strip: the "talkies" (1922).

These tentative developments in the *audiovisual* or bidimensional presentations in the early part of this century have today become virtually standard for much entertainment (cinema, television, MTV, video and camcorder technologies, etc.). In short, part of the ordinary lifeworld in technologically developed countries is the presence of many audiovisual media. Entertainment and communication have made this mode of bidimensional presentation a standard (and within this context, science education, documentaries, and other forms of canned presentations are also audiovisual).

Attempts to go beyond the audiovisual were also made—"smellovision," seat-shaking techniques in some movie theaters (tactile-kinesthetic), and other multisensory effects—but most were not successful. But the current, highly publicized attempts at the multisensory are clearly apparent in the developments of *virtual reality* (VR) technologies. And by way of historical leap, it is these to which I shall turn for speculation concerning going "beyond vision."

At the same time, even a sober speculation must qualify itself by the recognition that, simultaneously, VR technologies are still very "primitive" and also highly "hyped." What I intend to investigate by way of imaginative experience experiments are examples of some of the best VR technologies for the possible implications for scientific practice.

VR, in the sense in which I wish to investigate it, perhaps attains its current best use in *simulating learning or new experience problems within controlled contexts.* The origins of simulation go back to World War II and the development of the Link Trainer (1929). The "problem" for which the Link was developed related to the training of air force pilots, particularly fighter pilots. In the early days of World War I, the air force discovered that losses of fighter pilots were staggeringly high—until they somehow survived the first five missions. After that, the loss percentage dropped radically. Extended practice training in dogfights with more experienced pilots, while likely to produce the same or better results, was limited by the need to train new pilots quickly and to release experienced pilots for combat in this early air warfare. Following World War I, the Link Trainer was to fill this gap: a simulated cockpit was built, with "virtual" components such as (1) visual display before the cockpit screen showing the action of the enemy fighters, (2) controls linked to mechanized motions to mimic the twists and turns of an actual flight, and even (3) projected sounds of the action to be taken. This was a "quasi experience" of the dogfight scenario, used for quicker and yet "realistic" training, and it produced close to the desired results.

Since that time, training simulation has become much more sophisticated and used in more than military applications. Highly experienced pilots for commercial airlines undergo simulation testing (which entails crisis situations not likely to be actually experienced, such as engine failures, wind sheer phenomena, etc.) which is "realistic" enough that the pilots undergoing these tests report intensive anxiety and sweating responses—even though the experience is also clearly understood intellectually to be "merely" a VR experience. As with the earlier Link, these simulators include audiovisual plus tactile-kinesthetic dimensions. These experiences *within* the constructed "world" of the simulator are thus much closer to "whole body" experiences.

Here, as with visualism, whole body simulation calls for whole body "isomorphism." In recent simulator machines, for example, those which now train Lynx helicoper pilots, testers must see to it that the effects mimic the actual Lynx. In the early development of the simulator, a tester reported that "It flies like a pregnant duck," and only after much program adjustment could the simulator be accepted.[4]

Among the sciences, the adoption of VR simulations has expanded and now extends to both military and medical applications. Today VR is thought to be particularly useful for training surgeons, instrument operators, and technicians and pretraining for highly technologized operations within medical practice.

With another speculative leap, I would project that what is occurring within these contexts is a different version of what I have been calling a "Latourean laboratory." The VR version of the Latourean laboratory, however, is not necessarily primarily and certainly not solely focused upon producing a visual result. It is, rather, focused upon close to a whole body action and thus upon virtual interventions and manipulations with the objects of VR.

Still, many features of a laboratory remain constant: the VR context is a *closed context* and is flexible only within the limits of the program and the actually closed spatial-temporality which defines the VR context. Playing in VR contexts is analogous to playing any other kind of computer game—and the learning curve very much follows that of sophisticated video games and simulators in game contexts. At first there is fascination and a greater illusion of "real" action and challenge. But as levels are transcended and repeat scenarios show themselves again and again, eventually boredom and loss of interest can occur. Indeed, I would argue from my own experience (so far) that while there is a greater sense of whole body engagement, the "reality" experienced within VR contexts actually presents one with a stronger sense of *irreality* than found in the use of simpler and less multisensory technologies. (Compare the bodily adjustment nuances in learning to use a new pair of eyeglasses with the same adjustments in bodily motion "inside" the face-sucker mask and glove combinations of VR games.)

As a technique for teaching new skills—particularly those which themselves mimic the VR context—there is no doubt that such mediating technologies are useful and even—in medicine—theoretically could result in therapeutic improvement. The contemporary surgeon who practices microsurgery (laser surgery, endoscopy, etc.), by manipulating at a distance while watching a moniter, can use a good deal of VR training as the technology develops. He or she is also, in practice, much closer in the use of bodily motions to the skilled video game player than to the older hands-on, open-body surgeon.

Not everyone is happy, however, with this change of surgical context. Edward Tenner notes that such operations, in laparoscopy for example, have reductive effects as well:

> A television image is narrower, grainier, and of course flatter than the immediate if messy sight of an open abdomen. Surgeons can't feel internal organs directly with the fingers. To perform what critics have called "Nintendo sugery" requires skills different from those for traditional procedures—different enough that some surgeons with the right spatial-motor abilities are proficient after a few supervised operations while others are said to need dozens or even hundreds.[5]

This situation mimics that of a movie, *The Last Starfighter,* in which an ace video game player (the young hero) is tapped by aliens who need him to fly missions in another galaxy since he has learned the skills in the video parlor. In "RL" (real life), Tenner claims that complications arising from laparoscopy still run up to five times higher than traditional procedures (43).

This VR version of a Latourean laboratory, moreover, comes even closer to Robert Crease's theater production metaphor for science experimentation.[6] One "enters" quite literally and bodily the VR context. It is closed off (controlled) from the open ordinary world, and it displays its own "world" in terms of the various decison trees which are built into the program. Much of the hype concerning VR—can virtual reality *replace* RL?—revolves around the old theatrical distinction regarding suspended disbelief. The bumpkin who gets up from the audience to punch out the villain in a melodrama is projected in VR hype to be anyone whatsoever. (Anyone donning mask and gloves, who then thinks he or she is "entering reality," must be just such a bumpkin.)

Returning, however, to VR in a scientific context, what beyond the skill-training capacities of simulation might be expected? The answer may be a parallel to *remote sensing* in the form of *remote action.* Within the laboratory there is action: the preparation of the experiment, the elaborate production, the calibration and manipulation, and the eventual carrying out of the experimental action. In the contemporary scientific situation remote sensing has attained the level of a high art—could the same occur with respect to remote action?

Return to the example of Sojourner (the small robot exploring the Martian surface): this now "cheap" robot (only $227 million, the equivalent of two full years of total funding for the poor, beleaguered NEA!) is now remote controlled by signals conveying discrete commands from Earth. Its bumbling, if often successful, movements, under a future possible VR action-at-a-distance scenario, could have an Earthbound pilot making control movements not unlike the surgical operations-at-a-distance now envisioned by the military as an application of medical VR practice. And, were there sufficient feedback mechanisms to make the "feel" right, we would be on the verge of remote action embodied in robotics for solar system exploration.

This is not a "whole body hermeneutics" since it could well be that whatever actions taken on Mars would still be reported and used as "proofs" within scientific visual hermeneutics and could still be simply visual products. Yet the intriguing question arises—with a more complete interaction with things, does the traditional "distance" of the visual continue to hold? Or is there another secret liaison between science and the visual, one which interprets vision itself as an "objectifying" gaze which is not to be breached in a more bodily "dance" with things?

Or . . . is there a possibility for a groundswell shift toward a whole body science action? I will conclude my speculations with two examples of forefront technologies which pertain to bodily action. First, another military example (because the military often develops such frontier technologies because of favored funding and support, not because it is "scientific") from contemporary piloting: supersonic fighters became, in the last few decades, one of the most complex pieces of military, single-piloted, technologies. Instrument panels were replete with dials, auditory alarms warned of critical levels of machinery overload, remote sensing informed of incoming missiles, and so on. The problem which emerged was "information overload," a not unfamiliar problem of hi-tech life. Pilots, in high-speed dogfights or evading heat-seeking missiles, tended to "blank out" with this overload and often reverted to "seat of the pants" flying. Too much, too quick, too *cognitive.* One form of technical development to which the military is often sensitive is the use of ergonomics (technologies "friendly" to human, here bodily, use). Recognizing the need to capture the actions of a full body response, one device became favored for combat situations: the *heads-up display* helmet.

This device is a *composite,* but ergonomic, device which combines aspects of previously noted visualism with fast body response. The helmet's visor has projected upon it a grid with target-centering markers and a lighted identifier for the proper range of missile release. The pilot can use "normal" head motion in flying and yet have the calculations done and projected upon the visor. Here perceptual and actional immediate bodily Gestalts combine with the graph (textlike) visuals of a computer-generated target calculator in composite fashion. Enhanced, simultaneously, are bodily and visually (hermeneutic) actions. The superhuman speeds of the fighter "return" in part to the bodily capacities of the human pilot.

The second composite situation which combines the best visualist techniques with bodily action, but in a slower time, is the use of a complex, contemporary set of navigational instruments for the captain of a small sailing vessel. The situation is one in which wind and fog have combined and finding the entrance path to a rock-strewn harbor is essential. A GPS (global positing satellite device) displays either the set of numbers (coordinates), or an overhead chart (oceanographic map) display, or a display of lines converging into the harbor entrance (à la Renaissance perspective), or, possibly, the alternation between the displays. This device "pierces" the fog and clouds and allows close-up vision, combining as above perceptual isomorphic with graphlike features. One could add today a forward and side-scanning sonar device which again displays, this time, the *underwater* topography (to avoid reefs and rocks and shallows), making what is under the vessel transparent. Thus, while the captain must use his or her kinesthetic abilities to steer in the waves and through the gusts, the visual displays clearly indicate what could not be clearly seen in the conditions given.

Admittedly, these are not "science" examples, but they do indicate how composite (visual) techniques can be combined with whole body action in a live situation. If VR projections for similar bodily action can be perfected for remote action, then at the least the possibilities of investigation are made even more open than we have seen to date. Latour's understanding of an instrument (producing a visual result) no matter how expensive, complex, or layered applies here as well to whole body action. The instrument complex can be *embodied* if the qualifications are such that it is made ergonomically, and the body-perceptual capacities of humans are understood and taken into account. What new science could emerge from such a technology complex, or what new hermeneutic enhancements could be invented, I leave to my readers.

Afterword

In this attempt to expand hermeneutics, I have primarily demonstrated that the practices of the sciences include strongly hermeneutic dimensions. Beginning with early Modern science, but accelerating in late Modern science, those hermeneutics have been focally visualist but are interpretive activities nonetheless. In effect—although I have not and will not claim that scientific activity is either reducible to or always dominantly an interpretive process—this expansion of hermeneutics does carry some interesting implications for the understanding of the sciences.

Despite today's "science wars" in which some science conservatives wish to maintain a view of science which continues to assert "universality," "absoluteness," and culture-free "truth values," no serious group of interpreters of science maintain this late-nineteenth- and early-twentieth-century view. Philosophers of science, probably the most conservative of science interpreters, now are all "fallibilists."[1] The social studies of science including historians, sociologists, and anthropologists, are more radical. And although what is called "social constructionism" does not often actually hold that scientific truths are merely, or reducible to, the social processes which produce them, the embeddedness of science within society and culture is clearly part of a consensus about the sciences in the late twentieth century.[2] Feminists take the even further step of seeing science practice as deeply embedded in and containing gender perspectives, and most recently, gender perspectives which also echo multicultural and postcolonial aspects.[3]

A hermeneutic reframing of the sciences is a compatible perspective within this milieu of fallibilist late-twentieth-century thinking. And it has some advantages if pushed to its limits—what if the sciences were thought of as deeply or primarily hermeneutic activities?

First, if the sciences do contain deep interpretive activities, this insight places the sciences much more closely to the other intellectual practices known as the humanities and the social or human sciences than was thought to be the case earlier in this century. But, second, if the sciences are deeply interpretive and seen to be such, then as yet another human interpretive activity, much of the older mythology about the sciences should be diminished. No one (except fundamentalists who claim "literal" interpretations of certain texts) holds that in hermeneutics, any single interpretation can be absolute, final, or even universal. Instead, interpretations are ongoing, contestable, and changing: *just as in the sciences as portrayed in late-twentieth-century science studies.* And, if the hermeneutics is critical, then one can still claim that there are better and worse interpretations, more and less critical interpretations, insightful and blind interpretations—but no final or perfect ones. A hermeneutic reframing of the understanding of science thus not only fits into the fallibilist mentality of the present, but also avoids all of the pre-Kuhnian accounts of science as a simplistic linear accumulation of knowledge about Nature.

Ironically, this does return us to where we began, with echoes about the variants upon hermeneutics which were alluded to in the earlier religious orientation of interpretation theory. Were the conservatives of science forced to choose what variety of hermeneutic theory they would espouse, the answer would probably be one which—not far from the fundamentalists—would seek some original truth, the original "text" as given to Moses or the church fathers. And even if today's science fundamentalists would eschew claims about having such truths in hand, they would maintain that one could ever more closely approach such truths. But, in the history of hermeneutics, this choice would be a precritical one, today thoroughly outdated and archaic.

Another choice, rarely suggested but implicit as some have shown,[4] would be to take a Judaic route. Here, while still holding to the existence of an original text (the Torah), interpretation proceeds in many layers, continuously, and in a process of never ending contestation through the types of commentaries continuously developed in the living tradition. In some ways this process does actually look more analogous to science history than the account of that same history in the positivist accounts. The hidden irony here is that this mode of endless, indeterminate, and yet creative and critical hermeneutic is that which is also embedded in a postmodern *deconstructive* style of interpretation.

The third implication, which again moves a hermeneutic reframing of science understanding toward the late-twentieth-century science studies consensus, is that if the sciences may be deeply understood as a special type of interpretive activity, then the "special" privileges given to the sciences must appear to be more *political* than epistemological in nature. That is clearly the import of much of the science studies reframing process at present. The implicit understanding of this implication is one which lies behind so many of the dynamics of the "science wars." Indeed, the explicit worries which crop up in the controversies often point to the need to maintain the "faith" and "belief" in science which then keeps its funding sources flowing.[5]

These implications are, of course, already understood in different ways in the late-twentieth-century attempts to understand the sciences. A hermeneutic reframing of science understanding is not neutral and contains its own dangers. But it also points to a kind of truth-seeking which is not bound, and has never been bound, to the limits of a Modernist epistemology. Such a hermeneutic reframing, however, also has its own limits. It, too, is perspectival. It does not deal well with the social-political changes which also characterize late Modern science, that is, "Big Science," the institutionalization of so much of the research process within governmental, industrial, and other now often supranational institutions. For such analysis, other complementary perspectives are needed, such as those provided by Critical Theory, the feminist interpretations of the sciences, and the other social studies of science. This recognition and acceptance of limitation, however, is in keeping with the pragmatic spirit of contemporary hermeneutics as well. There is no "last" or "final" word, but there are more revealing and less revealing words.

I end with a personal confession—the more I have studied the visual hermeneutics of the sciences, the more I have become fascinated with what can be "seen." And this new set of visions, following the trajectories of the ever more micro- and macro-dimensions of what is "out there" enhances and does not in any way diminish the sense of awe which motivates the sciences at their best. If the sciences are fallible, contingent, and subject to change, then they also sit more easily within what we already know about the human condition.

Notes

Introduction

1. The three articles which precede part 4 include "Perceptual Reasoning" and "Expanding Hermeneutics," which both appear in *Hermeneutics and Science,* ed. Marta Feher, Olga Kiss, and Laszlo Ropolyi (Dordrecht, 1998). The third is "Thingly Hermeneutics/Technoconstructions" in *Hermeneutics and Natural Science,* ed. Robert Crease and Robert Scharft (Dordrecht, 1997).

2. Paul Forman's review of *The Flight from Science and Reason* is in *Science* 276 (2 May 1997): 750, and the letters responding are in *Science* 276 (27 June 1997): 1952.

Chapter 1

1. Although the outline I have followed to this point is fairly familiar, I would refer the reader to a more detailed account of origins in Richard Palmer, *Hermeneutics: Interpretation Theory in Schleiermacher, Dilthey, Heidegger and Gadamer* (Evanston, 1969).

2. I consider my book *Listening and Voice: A Phenomenology of Sound* (Athens, Ohio, 1976) to be a continuation of the hermeneutic tradition in this direction.

3. Palmer, *Hermeneutics,* p. 15. Throughout, citations in parentheses refer to the previously cited work.

Chapter 2

1. Ricoeur, *Husserl* (Evanston, 1967), p. 210.

2. Maurice Merleau-Ponty, "What Is Phenomenology?" *Cross Currents* 4 (winter 1956): 65.

3. Merleau-Ponty, "The Primacy of Perception," in *Existential Phenomenology* (London, 1967), p. 41.

4. There is even some question about a possible hermeneutic direction in the late Merleau-Ponty; see *The Visible and the Invisible* (Evanston, 1968).

5. In certain respects, Heidegger is actually closer to a philosophic version of "symbolic interactionism" (Mead) than to existentialism.

6. Martin Heidegger, *Being and Time* (New York, 1962), pp. 61–62.
7. Paul Ricouer, *De L'Interpretation* (Paris, 1969), pp. 13–14.

Chapter 3

1. Carl Mitchum, *Thinking through Technology* (Chicago, 1994).

2. Don Ihde, *Technics and Praxis: A Philosophy of Technology* (Boston, 1979). Friedrich Rapp actually published his *Analytical Philosophy of Technology* in German in 1978. It did not appear in English, however, until 1981 in the same series as *Technics and Praxis.*

3. See *Instrumental Realism: The Interface between Philosophy of Science and Philosophy of Technology* (Bloomington, Ind., 1991).

4. Hubert Dreyfus, *What Computers Can't Do* (New York, 1972).

5. Patrick Heelan, "Horizon, Objectivity and Reality in the Physical Sciences," *International Philosophical Quarterly* 7 (1967): 375–412. See also *Space Perception and the Philosophy of Science* (Berkeley, 1983).

6. Human-produced spears have recently been found in northern Germany dated to 400,000 B.P. (There is some question, however, concerning thermoluminoscopic dating techniques.)

7. Bruno Latour, *We Were Never Modern* (Cambridge, Mass., 1993); see chapter 1.

Chapter 4

1. Edmund Husserl, *The Crisis in European Sciences and Transcendental Phenomenology,* trans. David Carr (Evanston, 1970), pp. 29–30.

2. Martin Heidegger, *The Question Concerning Technology,* trans. William Lovitt (New York, 1977), p. 3.

3. Martin Heidegger, "The Origin of the Work of Art," in *Poetry, Language and Thought,* trans. Albert Hofstadter (New York, 1971), p. 42.

4. J. Donald Hughes, *Ecology in Ancient Civilizations* (Albuquerque, 1975), p. 1.

5. L. Sprague de Camp, *The Ancient Engineers* (Garden City, N.Y., 1963), p. 93.

6. Heidegger, *The Question Concerning Technology,* p. 129.

7. Martin Heidegger, *Being and Time,* trans. John Macquarrie and Edward Robinson (New York, 1962), p. 207.

Chapter 5

1. Maurice Merleau-Ponty, *Phenomenology of Perception,* trans. Colin Smith (London, 1962), p. 179.

2. Ibid., p. 193.

3. Maurice Merleau-Ponty, "The Primacy of Perception," trans. James M. Edie, in *Existential Phenomenology* (London, 1967), p. 41.

4. *Phenomenology of Perception,* p. 177.

5. Maurice Merleau-Ponty, *Signs,* trans. Richard C. McCleary (Evanston, 1964), p. 43.

6. *Phenomenology of Perception,* p. 184.

7. *Signs,* p. 39.

8. *Phenomenology of Perception,* p. 188.

9. *Signs,* p. 42.

10. *Phenomenology of Perception,* p. 195.

11. Maurice Merleau-Ponty, *The Visible and the Invisible,* trans. Alphonso Lingis (Evanston, 1968), p. 103.

12. "The Primacy of Perception," pp. 42, 49.

13. *Phenomenology of Perception,* p. 197.

14. *The Visible and the Invisible,* p. 212.

15. "In a sense the whole of philosophy, as Husserl says, consists in restoring a power to signify, a birth of meaning, or a wild meaning an expression of experience by experience, which in particular clarifies the special domain of language. And in a sense, as Valéry said, language is everything, since it is the very voice of the things, the waves, and the forests. And what we have to understand is that there is no dialectical reversal from one of these views to the other; we do not have to reassemble them into a synthesis: they are two aspects of the reversibility which is the ultimate truth" (*The Visible and the Invisible,* p. 155).

Chapter 6

1. Paul Ricoeur's bibliography is among the largest of contemporary, forefront philosophers. The list of systematic books which undertake large projects include the trilogy revolving around the "philosophy of the will" in the 1950s (*Freedom and Nature, Fallible Man, Symbolism of Evil*) to the most recent trilogy on narrative (*Time and Narrative,* 3 volumes) in the 1980s, interspersed by regularly appearing single volumes. Alongside these larger, systematic works are regularly appearing books which collect his primary articles. In virtually every one of these collections are to be found essays upon hermeneutics, its history, and its relation to phenomenology. Of these, *The Conflict of Interpretation* (French ed., 1969; English ed., 1974), *Hermeneutics and the Human Sciences* (French and English eds., 1981), and *From Text to Action* (French ed., 1986; English ed., 1991) are prominent.

2. Ricoeur's discussions of Schleiermacher and Dilthey are to be found scattered through virtually every discussion of his on hermeneutics. See especially the essays referred to in note 1, but also see my "Interpreting Hermeneutics" (chapter 1).

3. My *Hermeneutic Phenomenology: The Philosophy of Paul Ricoeur* (Evanston, 1971) contains a detailed analysis of Ricoeur's dialectical approach in his early period (French publications through 1969). This work was the first English-language, systematic interpretation of his (early) corpus. I have found that once taken as a "scholar" of some major figure, there follows a fated series of invitations to update one's earlier work. That has also been my occasional fate, and these

updates on Ricoeur may be found in my introduction to the English translation of *The Conflict of Interpretations* (Evanston, 1974), "Interpreting Hermeneutics (chapter 1), "Variation and Boundary: A Problem in Ricoeur's Phenomenology," in my *Consequences of Phenomenology* (Albany, 1986), and "Text and the New Hermeneutics," in *On Paul Ricoeur: Narrative and Interpretation* (New York, 1991).

4. Ricoeur's most explicit discussion of a rejection of Hegel's approach to history may be found in *Time and Narrative*, vol. 3 (Chicago, 1988); see especially chapters 8 and 9.

Chapter 7

1. Jacques Derrida, *Speech and Phenomena*, trans. David Allison (Evanston, 1973), p. 103.
2. See Michel Foucault, *Discipline and Punish* (New York, 1979).
3. See Iris Young, *Throwing like a Girl and Other Essays* (Bloomington, Ind., 1990), chapter 11.
4. Jean Baudrillard, *The Gulf War Never Happened* (Bloomington, Ind., 1995).
5. Evelyn Fox-Keller *Reflections on Gender and Science* (New Haven, 1985), p. 10.
6. See my *Technology and the Lifeworld* (Bloomington, Ind., 1990) and *Instrumental Realism* (Bloomington, Ind., 1991).
7. Robert Crease, *Science* 261 (30 July 1993): 555.

Chapter 8

1. Derek Parfit, *Reasons and Persons* (Oxford, 1986), p. 199.
2. Paul Ricoeur, *Oneself as Another* (Chicago, 1992), p. 113.
3. Charles Reagan, *Paul Ricoeur: His Life and His Work* (Chicago, 1996).
4. Ricoeur, *Oneself as Another*, p. 136.
5. Parfit, *Reasons and Persons*, p. 200.
6. Ricoeur, *Oneself as Another*, p. 150.
7. Ibid., p. 150.
8. Don Ihde, *Technology and the Lifeworld* (Bloomington, Ind., 1990), p. 75.
9. Ricoeur, *Oneself as Another*, p. 150.
10. Ihde, *Technology and the Lifeworld*, p. 75.
11. Ricoeur, *Oneself as Another*, p. 325.
12. The data I cite here are constructed from information provided by colleagues at Stony Brook in astronomy, physiology, and genetics.
13. Ricoeur, *Oneself as Another*, p. 135.
14. Ibid., p. 197.

Chapter 9

1. Robert Nozick's *Philosophical Explanations* was the year's contender. But if my small sample is indicative, whereas Rorty was read, Nozick's tome rarely was finished by readers.

2. I continue the convention of the introduction here with ACE standing for the American Continental Establishement and AE the Analytic Establishment.

3. Richard Rorty, *Philosophy and the Mirror of Nature* (Princeton, 1979), p. xiii.

4. Others are also aware of the need to recharacterize the practice of analytic philosophers, as in Moulton's use of a legal practice.

5. I have long contended that Husserl must be read through. His heuristic discourses on method are attempts, after he has seen something difficult to see, to tell others how to do it. By adopting extant terminologies and then reversing or radically changing their meanings, his work is almost metaphorical. Heidegger, I would contend, must be read literally.

6. To term these Gestalts and imply Husserl used them is a bit anachronistic since the Gestaltists were aware of, and in some cases were students of, Husserl.

7. Rorty, *Philosophy and the Mirror of Nature,* p. xiii.

8. Privately circulated, this list apparently came out of one of the Dreyfus summer programs.

9. A dissertation by Gary Aylesworth traces both the Wittgensteinian and Heideggerian directions carefully (Stony Brook, 1984).

10. Rorty, *Philosophy and the Mirror of Nature,* p. 394.

11. See my "Phenomenology and the Later Heidegger," which shows the way in which phenomenology functions in his later works (in *Existential Technics,* Albany, 1983).

12. Don Ihde, "Phenomenology and Deconstructive Strategy," *Semiotica* 41 (1982), pp. 5–24 (reprinted in *Existential Technics*).

13. Chapter 2 of this collection (*Consequences of Phenomenology*) goes into more detail on Merleau-Ponty and Foucault.

Chapter 10

1. Langdon Winner, "Paths of Technolopolis" (forerunner manuscript to *The Whale and the Reactor* [Chicago, 1984]), p. 3.

2. Valerie Flint, *The Imaginative Landscape of Christopher Columbus* (Princeton, 1992).

3. Bruno Latour, *Science in Action* (Harvard, 1987), p. 61.

4. Raphael Sassower, *Knowledge without Expertise: On the Status of Scientists* (Albany, 1993), p. 65.

5. *Science,* 23 February 1996: 1050–52.

6. *Science,* 5 January 1996: 35.

7. "A Partially True Story," *Scientific American* 272 (April 1993).

8. Don Ihde, *Philosophy of Technology* (New York, 1993).

Chapter 11

1. Jeffrey Burton Russell, *Inventing the Flat Earth: Columbus and Modern Historians* (New York, 1991).

2. Stephen Fuller, "The Philosophy of Science since Kuhn," *Choice* (December 1989): 595.

3. Paul Feyerabend, *Against Method* (London, 1975).

4. Thomas Kuhn, *The Structure of Scientific Revolutions* (Chicago, 1962).

5. See, e.g., David Bloor, *Knowledge and Social Imagery* (London, 1977); Karin Knorr-Cetina, *The Manufacture of Knowledge* (1981); Bruno Latour and Steve Woolgar, *Laboratory Life* (Beverly Hills, 1979); and Andrew Pickering, *Constructing Quarks* (Chicago, 1984).

6. See *The Social Dimensions of Science* (Notre Dame, 1992), a collection edited by Ernest McMullen in an attempt to modify the more radical conclusions of "social constructionism."

7. Peter Galison is an excellent example of such a shift; see his *How Experiments End* (Chicago, 1987).

8. See, e.g., Sandra Harding, *The Science Question in Feminism* (Ithaca, N.Y., 1986); Evelyn Fox-Keller, *Reflections on Gender and Science* (New Haven, Conn., 1985); and Donna Haraway, *Simians, Cyborgs, and Women* (New York, 1990).

9. Examples of the work of these philosophers appear in the recent *Hermeneutics and the Natural Sciences,* ed. R. Crease and R. Scharff (Dordrecht, 1997). Also see Robert Crease, *The Play of Nature* (Bloomington, Ind., 1993).

10. Don Ihde, *Instrumental Realism* (Bloomington, Ind., 1991).

11. Joseph Rouse, *Knowledge and Power: Towards a Political Philosophy of Science* (Ithaca, 1987); Bruno Latour, *Science in Action* (Cambridge, Mass., 1987).

12. Rouse, *Knowledge and Power,* p. 41.

13. Latour, *Science in Action,* pp. 67–68.

14. Don Ihde, *Technics and Praxis: A Philosophy of Technology* (Boston, 1979).

Chapter 12

1. See my *Listening and Voice: A Phenomenology of Sound* (Athens, Ohio, 1976).

2. Husserl distinguishes between the lifeworld and the "world" of science: "The life-world is the world that is constantly pregiven, valid constantly and in advance as existing. . . . [E]very science presupposes the life-world; as purposeful structures they are *contrasted* with the life-world, which was always and contines to be 'of its own accord' "; Husserl, *The Crisis of European Sciences and Transcendental Phenomenology,* trans. David Carr (Evanston, 1970), p. 382.

3. Merleau Ponty continues the distinction made by Husserl: "The whole universe of science is built upon the world as directly experienced. . . . Science has not and never will have, by its nature, the same signifiance qua form of being as the world which we perceive. . . . [S]cientific points of view, according to which my existence is a moment of the world's, are always both naive and at the same time dishonest, because they take for granted, without explicitly mentioning it, the other point of view . . . through which from the outset a world forms itself round me and begins to exist for me"; *Phenomenology of Perception,* trans. Colin Smith (London, 1962), pp. viii–ix.

4. This is a sustained thesis in my *Technology and the Lifeworld* (Bloomington, Ind., 1990).

5. Harold I. Brown, "Galileo on the Telescope and the Eye," *Journal of the History of Ideas* 45 (1985): 487.

6. Ibid., p. 489.

7. Kuhn cites numerous examples of this problem, placing it in the context of changed perceptions—he claims that one Gestalt shift concerning "chaff particles" in the seventeenth century becomes "electrostatic repulsion" in the nineteenth, but he admits that "electrostatic repulsion was not seen as such until Hauksbee's large-scale apparatus had greatly magnified its effects"; Thomas Kuhn, *The Structure of Scientific Revolutions* (Chicago, 1962), p. 117.

8. Galison argues that experiments do not end—rather, there are repeated endings and refinements which often result in a general agreement that phenomena are established when effects would not "go away"; see *How Experiments End* (Chicago, 1987), p. 237.

9. My son, Mark, theorized that the claims were made by people who needed glasses. He noted that when he took his own glasses off, the Moon did appear smooth, so maybe the church fathers were the old men who could write—or, perhaps, the dark and light contrasts reflected Earth features?

10. Harding's reference is to Ivan van Sertima, *Blacks in Science: Ancient and Modern* (New Brunswick, N.J., 1986).

11. See Hubert Dreyfus, *What Computers Can't Do* (Cambridge, Mass., 1993), chapter 7.

12. A more complete discussion of Leonardo's transformation of vision can be found in chapter 1 of my *Postphenomenology* (Evanston, 1993).

13. See Bettyann Holtzmann Kevles, *Naked to the Bone: Medical Imaging in the Twentieth Century* (New Brunswick, N.J., 1997).

14. See my *Experimental Phenomenology* (Albany, 1986), pp. 128–29.

15. Kuhn, *Structure of Scientific Revolutions,* pp. 115–16.

16. See Lee W. Bailey, "Skull's Darkroom: The *Camera Obscura* and Subjectivity," in *Philosophy of Technology* (Dordrecht, 1989).

17. See both Jon Darius, *Beyond Vision* (Oxford, 1984), pp. 34–35, and Peter Pollack, *The Picture History of Photography* (New York, 1977), pp. 65–67.

18. Darius, *Beyond Vision,* pp. 34–35.

19. I have traced some interesting cross-cultural aspects of the imaging of others in pre- compared to postphotographic contexts; see chapter 4 of *Postphenomenology* (Evanston, 1993).

20. See Kevles, *Naked to the Bone,* pp. 3–20.

21. Ian Hacking, *Representing and Intervening* (Cambridge, 1983), p. 193.

22. Ibid.; see the chapter on "Microscopes," pp. 186–209.

23. Oscar Ogg, *The 26 Letters* (New York, 1967), p. 78.

24. These patterns are used by speech pathologists, for example, to show speakers how what they are saying does not, in fact, correspond to the standard form of a native language.

Chapter 13

1. Nigel Henbest and Michael Marten, *The New Astronomy,* 2d ed. (Cambridge, 1996), p. 6.

2. Bettyann Holtzmann Kevles, *Naked to the Bone: Medical Imaging in the Twentieth Century* (New Brunswick, N.J., 1997), p. 20.

3. Kevles provides a time chart, paralleling the various developments in the multiple imaging instrumentation.

4. *Science* 274 (13 December 1996): 1870–73.

5. *Science* 276 (30 May 1997): 1331–34.

6. *Scientific American* 276 (August 1997): 46–47.

7. *Science* 243 (31 March 1989): 1663.

8. *Science* 277 (11 July 1997): 176–78.

9. *Scientific American* 276 (August 1997): 61–65.

10. The visit to the Cholula pyramid and conversations with anthropologists occurred during the 9th International Conference of the Society for Philosophy and Technology, November 1996.

11. Postmodernism is more thoroughly discussed in *Postphenomenology* (Evanston, 1993).

12. *Micropaedeia, Encyclopaedia Britannica,* 15th ed. (Chicago, 1994), vol. 10, p. 514.

13. Kevles, *Naked to the Bone,* p. 15.

14. I refer to Hacking's "if you can spray them then they are real" in *Representing and Intervening* (Cambridge, 1983), p. 23.

15. *Science* 276 (27 June 1997): 1994. This rhetoric is an example of the more-than-neutral language often employed by science reporting.

Chapter 14

1. Bettyann Holtzmann Kevles, *Naked to the Bone: Medical Imaging in the Twentieth Century* (New Brunswick, N.J., 1997), p. 13.

2. Zermeiz is well known in biological circles for his pioneering work in ecology and evolution.

3. *Scientific American* 276 (June 1997): 84–87.

4. Internet printout from Lynx site provided by Marshall Spector.

5. Edward Tenner, *Why Things Bite Back: Technology and the Revenge of Unintended Consequences* (New York, 1996), pp. 43–44.

6. See Robert Crease, *The Play of Nature* (Bloomington, Ind., 1993), especially chapter 4.

Afterword

1. Larry Laudan, perhaps as good a spokesman for these positions as any, repeated affirms the fallibilist consensus for all the interlocuters in his *Science and Relativism* (Chicago, 1990).

2. The moderate position is summarized by a collection edited by Ernan McMullen, *The Social Dimensions of Science* (Notre Dame, 1992). Similarly, as evidence of the accommodation within science, the course currently taught and funded by the NSF called the Social Dimensions of Science (SCI 242) adopts this position and is used to enhance the education of science majors.

3. See Sandra Harding, *Is Science Multicultural? Postcolonialism, Feminisms, and Epistemologies* (Bloomington, Ind., 1998).

4. I have suggested this parallel in my "Phenomenology and Deconstructive Strategy," in *Existential Technics* (Albany, 1983), pp. 159–81.

5. I referred to the earlier *Nature* controversy (1987) in my *Technology and the Lifeworld* (Bloomington, Ind., 1990), but the more recent controversies surrounding *Science* and the response to "postmodernist" threats to science echo the same fears.

Index